# Everything You Know About the Human Body Is Wrong

*This book is dedicated to the virtuous individuals
who have donated their bodies to medical science.*

First published in the United Kingdom in 2018 by Batsford
43 Great Ormond Street
London WC1N 3HZ
An imprint of Pavilion Books Company Ltd

ISBN: 9781849944311

A CIP catalogue record for this book is available from
the British Library.

25 24 23 22 21 20 19 18
10 9 8 7 6 5 4 3 2 1

Reproduction by Mission Productions, Hong Kong
Printed and bound by Imak Ofset, Turkey

This book can be ordered direct from the publisher at the
website www.pavilionbooks.com, or try your local bookshop.

Distributed in the United States and Canada by
Sterling Publishing Co., Inc.1166 Avenue of the Americas,
17th Floor, New York, NY 10036

# Everything You Know About the Human Body Is Wrong

Matt Brown

BATSFORD

# Contents

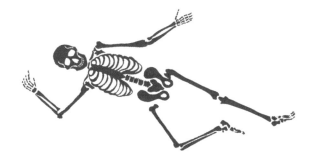

# Introduction

We spend every moment of life inhabiting our bodies. Nothing in the world is more familiar. Yet the human body is also very unfamiliar. Unusual circumstances aside, few of us ever see or touch the vast majority of our own bodies. It is a rare and unlucky person who gets to view his own femur or examine her own spleen.

The ancient Greeks had a saying: 'Know thyself'. Few of us come close. The inside of the body is a complicated, three-dimensional morass of organs, vessels, valves, tendons, muscles, fluids, bones, nerves, cartilage, nodes, sacs, flaps and countless other tissues beyond cognizance. You might spend years mastering the gross anatomy, and still have much to learn.

The human body is home to hundreds of species of bacteria, with many fungi, viruses and even animals along for the ride. Can anyone ever know the body at a cellular level? Scientists are only beginning to piece together the molecular machinery that underpins the whole enterprise. It would be a lesser task to memorize the whole of human literature than to carry out a complete survey of every physiological process in one human being.

Our bodies are marvellous, in the true meaning of that word. Nothing ever discovered or created comes close to the human brain in complexity or ability. The brain remains the most mysterious component of the body. We might be close to understanding how memories form, for example, but we are nowhere near appreciating how the physical memory encoded in neurons turns into a re-lived experience somewhere in our heads.

As such an intricate, complicated and still partly mysterious entity, the human body is a fertile breeding ground for myths and misunderstandings.

Unless you study it for a living, you've probably got very little idea how your gut works, how pain is transmitted from your fingers to the brain, or how a fetus develops in the womb. But because our bodies are the ultimate personal possessions, we all like to have working theories. Many old wives' tales still linger from simpler times, when nobody really knew what caused colds, cold sores and cancers.

This book is part of an ongoing series that examines and debunks the common myths of our daily lives. As with other volumes, the aim is not to ridicule or belittle, but to open the eyes to the infinite joy of knowledge. We learn most powerfully from our mistakes. A book exposing those mistakes should be more memorable than a straightforward tour of the body. And more fun.

Many of the entries in this book touch on issues of health and wellbeing. While I have, of course, striven for accuracy, any information herein should not be treated as a substitute for the medical advice of a doctor or health-care professional. Further, the reader is encouraged never to take statements at face value, including ones made here. Always dig if something sounds dodgy.

Let the nitpicking commence!

---

A NOTE ON SOURCES: Wherever relevant, I've checked my facts by consulting the scientific literature – research articles that have been carefully reviewed and published in respected journals. A book like this needs to strike a balance. To cross-reference every other sentence would be cumbersome and tedious, but to leave out all references would be negligent. I've therefore included references for information that is either particularly surprising, or for which the reader might enjoy probing a bit deeper than my own word limit allows. In such cases, I've cited the name of the first author and a 'doi' number. The doi, when typed into an Internet search bar, should lead you to an online version of the article (although it may be partially behind a paywall).

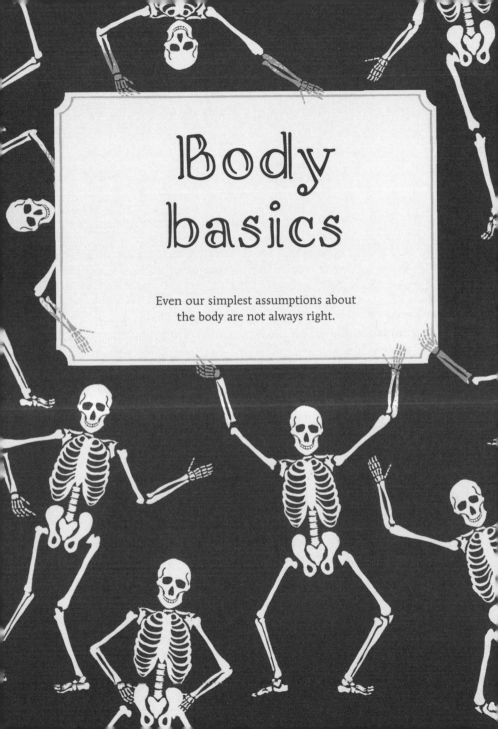

# Body basics

Even our simplest assumptions about
the body are not always right.

# We know everything about the human body

Imagine climbing on board an aeroplane that nobody understands. The pilot is sure which buttons to press and which sticks to pull, but she has no clue why the plane stays in the air. The ground crew are able to make repairs and know that they must fill the plane with kerosene, but they don't understand how this makes the plane move, let alone fly. Even the aircraft's designers don't get it – they rivet some wings and a tail onto a fuselage and it somehow takes off. It works every time, so everyone carries on flying.

This scenario is clearly ludicrous, but it's not far from the situation with anaesthetics.

For most of history, patients undergoing surgery had to endure unimaginable horror. Flesh was cut and bone was cleaved without any effective pain relief. Then, from the mid-19th century, anaesthetic gases became available. This changed everything. After a few decades of trial and error, the use of anaesthetics removed most of the terror from the surgeon's table. Speed was no longer of the essence, and patients had no capacity to writhe around. Those wielding the scalpel could do so under controlled conditions and survival rates rocketed accordingly. Modern anaesthetics are extremely safe (when administered by a professional), reliable and used millions of times each year.

Nobody knows how they work.

Clearly, a general anaesthetic must act on the brain in some way. The agent turns off consciousness, while leaving other brain activities intact. There are theories on how the chemicals in the anaesthetic interact with the brain, but nobody has yet set out a detailed mechanism. It goes to show just how much we still have to learn about the body that, after a century-and-a-half of use, anaesthetics retain this mystery.

The human body, it hardly needs saying, is complex. Modern imaging techniques have taken knowledge to levels unimaginable only a few generations ago. And yet there is still much to learn – not just at the level of molecules and biochemistry, but even in the realm of anatomy, which concerns itself with the body at larger scales.

In early 2017 newspapers around the world announced that the body has a new organ. For decades, doctors had counted 78 organs, but a new discovery had topped that up to 79*. How, after thousands of years of study, could we have missed a whole organ? The culprit is a part of the digestive system known as the mesentery. It connects the intestine to the stomach. This is a pretty important junction. Indeed, the mesentery has been known since antiquity, so in one sense it's a little disingenuous to repeat the newspaper headlines of 2017 and describe it as a newly discovered organ.

In another sense, it really is a startling find. Doctors had always assumed that the mesentery was made up of a few fragmentary pieces, but more

---

*FOOTNOTE: This tally was widely reported in the newspapers, probably inspired by Wikipedia. It is spurious. There is no officially agreed number of organs in the body. It depends on how you define an organ. A common description holds that an organ is a self-contained part of the body that performs a specific function or functions. The heart, say, is a stand-alone unit with the clear task of pumping blood around the body. But the definition is a bit wishy-washy. My big toe is also a discrete, self-contained part of my body with the sole function of keeping me on my feet – why does that not count as an organ? How about all the bones in the body? The tally of 78 or 79 organs makes for good copy in newspaper stories but is not supported by the anatomical community.

recent scrutiny shows it to be a single structure. Hence, it now looks set to join the other organs on an equal footing.

What other mysteries lie within? Enough to keep the Nobel Prize for Physiology and Medicine well stocked with winners for the foreseeable centuries. The brain remains the most obvious conundrum. Knowledge of that organ leaps forward every year, but nobody knows how memories, thoughts and emotions can arise, and be experienced, within its bounds. How can a few pounds of intelligent tofu, evolved over millennia to catch gazelle and forage for berries, also calculate the composition and gravity wells of distant stars – and be self-aware enough to contemplate this impressive ability?

Why are the vast majority of humans right-handed, independent of culture and background? Indeed, why do we have one dominant hand anyway? Why did humans evolve different blood types, or unique fingerprints, or the ability to blush (which can put us at a social disadvantage)? Do pheromones play a role in human sexual attraction? Why have we kept some clumps of body hair and not others? Why does childhood and adolescence last so long in humans compared with other primates? The list is as long as your arm. Indeed, why your arm should be much weaker than a chimp's when you have similar musculature is just another example.

# Humans have five senses

You might have seen or heard that humans have five senses, but if you sniff around, you'll get the feeling that this is just a taster. As well as sight, hearing, smell, touch and taste, our bodies have a whole motherboard of sensors that give us information about our environment.

Close your eyes and lean forward. Even without seeing, you know that your body is close to tipping over. The information is not wholly coming from your sense of touch. We all carry something analogous to a spirit level in our heads. The inner ear contains three loop-shaped canals filled with fluids. As the head tilts or rotates, the fluids shift position. This movement is registered by tiny hairs, which in turn send signals to the brain. The brain determines if corrective motion is needed and instructs the muscles accordingly. We call it our sense of balance, and it is mediated by the vestibular system. There's no reason not to count it as a sense.

Then we have proprioception. This is the 'where?' sense that reports back on the body's conformation – simplistically, the direction in which your arms and legs are pointing or moving. With eyes closed, most people can touch their nose or scratch their head. Walking in the dark is physically no harder than walking in the light, so long as you know the path is clear. Proprioception has its own sensors called – logically enough – proprioceptors. These are special sensory fibres within the muscles that give feedback to the brain.

We can add still further powers to the sensorium. Our sense of touch, for example, can be broken down into its different facets. Pain, pressure, itches, heat and cold all feel different on the skin. This is because each is registered by a different type of receptor, and we might therefore regard them as separate senses.

Arguments can be made for still further senses. If we consider proprioception to be a sense, then why not other internal signals such as hunger or thirst? Ultimately, it's down to how one defines a sense, but it's clear that the traditional five are only part of the story.

We may have more than five senses, but our sampling of reality is hopelessly limited. Take our sense of sight. Only a fraction of the light that enters our eyes will trigger photoreceptors. Here, 'fraction' is an understatement. The visible spectrum is less than one-thousandth of 1 per cent of the wavelengths that hit our eyes. We never evolved the ability to see microwaves, UV, infrared, or any other flavour of electromagnetism beyond the spectrum of colours*.

Perhaps one day humans will enhance our senses through genetic engineering or technological bolt-ons. Might we soon be able to 'see' radio waves, or watch as microwaves cook our food, or enjoy 360-degree vision? Maybe our perception will grow to encapsulate the experiences of other humans, or even machines.

Experiments have already been conducted into sensing group emotions. Algorithms assess and compile the thoughts of thousands of people, as expressed on social media. The output is sent to a tactile vest, which bulges and buzzes to convey the emotions. The wearer learns to interpret prods

---

*FOOTNOTE: Or, rather, what we perceive as colours. There is nothing intrinsically yellow about a banana, and the sky is not in itself blue. Objects reflect and refract light in different ways, which we observe as different colours – but only because the human brain has evolved to file things that way. Some animals don't experience any colour, others can perceive light at wavelengths beyond our abilities. The banana would look very different in either case. We can imagine a creature that perceives different wavelengths of light as clicks of different frequency, rather than a visual sensation. The quality of 'yellowness' is merely an interpretation of an input. It's all in the brain, not in the banana.

from the vest and thereby 'feel' the sentiments passing around online. Even the colloquial 'sixth sense' of mind reading could become a possibility through technology. Crude, mind-controlled artificial limbs are already a reality. As we learn more about the brain, it may become possible to hook up transceivers that send messages directly between minds. Then again, people have been saying that for generations.

Some animals already do have additional senses not open to humans. They experience a different reality we can only begin to imagine. Migratory birds, and even the humble robin, are able to 'see' the Earth's magnetic field. It helps them to navigate. Some snakes can see infrared. Other animals – such as sharks, platypuses and bees – can detect electric fields. Sharks use this talent to sense the tiny electric signals in the muscles of prey. Certain species of dolphin can do this too. Add on the echolocation used to find their way through murky water, and dolphins might just be the most sensitive mammals of all.

# Humans have ten fingers

You can tell if your child is going to be a pedant. The warning signs start early: when 'Can you count to ten on your fingers?' is answered by, 'No, but I can count to eight, because I only have eight fingers, and two thumbs,' then you know you've got a nitpicker in the family.

If you're the kind of person who likes to get competitive with small children, then you might counter the eight-finger assertion by showing the child a dictionary. Most allow both definitions. Depending on context, it's legitimate to say either that humans have eight fingers or ten fingers.

A pedantic uncle, listening in, could go still further. 'A given person might have ten (or eight) fingers,' he might say, 'but humans, on average, do not.' He would argue that many people lose fingers or hands during life, whereas very few are born with additional digits. Across the world population, this lowers the average number of fingers to something like 9.9. 'If you have ten fingers,' he'd conclude, 'then you have an above-average number of digits.'

'Aha!' cries a nitpicking aunt, who has been following with interest, 'But that only applies to the mean. Other measures of the average, such as median and mode, would still give an answer of ten fingers.'

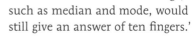

It is at this point that you back away from the conversation, wondering what you've started. You vow never to get your fingers – whether eight, 9.9 or ten of them – burned again.

# Most body heat is lost through the head

I must confess, as a gradually balding man, I've paid particular heed to this pearl of wisdom. Rarely will I venture outside on a cold day without my woolly headgear. So, I was most surprised to learn when researching this book, that there is no evidence to suggest the body loses most heat through the head. It's not even a particularly recent myth, but hats off to its tenacity.

Like all good fiction, the notion is underpinned with a convincing back story. The brain, it is said, uses so much energy that it radiates far more heat than any comparable area in the rest of the body. Hence, you should wear a hat in cold weather, or risk hypothermia.

This received wisdom was debunked in no lesser publication than the *British Medical Journal* – one of the world's leading scholarly journals*. The authors say that any uncovered part of the body would lose heat about as rapidly as any other. The head has no special radiative properties. It is simply the part of the body most often left uncovered. Few would go out in freezing conditions without a coat, but they might eschew a hat.

The myth has been traced back to an influential US Army survival manual of the 1970s, widely read by the general public, as well as service personnel. The guide suggests that up to 50 per cent of body heat is lost through the head – a claim repeated in modern versions of the manual. The assertion is thought to be based on flawed military research from the 1950s. Subjects were wrapped in survival suits and placed in Arctic conditions. They did indeed lose most body heat through their heads, but only because they weren't wearing hats. Had they been exposed in just their underwear, they would have lost heat proportionately from every nude surface.

The original military research is never properly cited and I've found it impossible to track down. It may itself be a myth, or a garbled truth. Even so, better documented research has arrived at the same conclusions. In one study (T. Pretorius et al., 2006, doi:10.1152/japplphysiol.01241.2005), researchers dunked hardy volunteers into cold water. Some were lowered up to their chins, others were completely submerged (wearing scuba gear). The experimenters could then measure the changes in body temperature. The head 'does not contribute relatively more than the rest of the body to surface heat loss,' they found.

In other words, if you were to stroll naked through the Arctic tundra, you would lose heat equally from all parts of your body. Perhaps 10 per cent of that loss would come from the head. But there is a catch, and one which throws a lifeline to the old myth. If you do wrap up warm but leave your head exposed, then your core temperature could suffer. The scalp contains more blood vessels than most other surface areas. Exposure cools the blood flowing through, which then travels down to the core. There's still every reason to wear that woolly hat in cold weather.

---

*FOOTNOTE: That said, it is reported in a whimsical Christmas-themed article, rather than a peer-reviewed write-up. Some scepticism is required.

# Veins are blue and arteries red

Some people think that human blood is blue. I had no idea that such a misconception existed until I started researching this book, but it seems to be a fairly common belief. The pseudoscientific reasoning goes like this. Anyone with light skin can see that veins are blue, just by looking at the back of their hand. The veins are blue because they contain blood that is blue. The blood only turns red when it is oxygenated by the lungs and pumped away through the arteries. We don't see blue blood when we cut ourselves, because it immediately oxidizes on contact with air, and turns red.

It sounds mildly convincing but just about every sentence in that chain of statements is flawed or wrong. Nobody has blue blood, not even the Queen. Blood does come in different shades of red – a bright crimson when oxygenated in the arteries, and a duller, darker colour when venous – but never, ever blue. Where does the idea come from?

Well, it's partly from the veins-in-the-back-of-the-hand thing. Were you to slice open your hand to take a look – please don't – you would find that all of your blood vessels are a pinkish-red, occasionally verging on maroon. They may appear blue from the outside, but this is a trick of the light. Remove the skin, which filters the light reflecting off the vein, and all is red. The colour comes from a molecule called haemoglobin. It is the iron atom at the centre of the haemoglobin which binds to oxygen, allowing it to be transported around the body. A common misconception – and one I was taught at school – would have us believe that the blood is red because of this oxidized iron in an aqueous environment (think rust, which forms when iron, air and water combine). Actually, the colour is chiefly caused by the wider structure of the haemoglobin, not the iron.

The myth of blue blood is further reinforced by anatomical diagrams and models. As a convention, these usually show arteries in red and veins in blue. It's a useful way of distinguishing these two very different types of blood vessel, but it doesn't match the physical reality.

Some animals do have blue blood, including spiders and crustaceans. Their blood is coloured by haemocyanin and copper rather than haemoglobin and iron. Certain worms have green blood. Strangest of all is *Chionodraco rastrospinosus*, the ocellated icefish of Antarctica, whose blood is transparent. These evolved to transport oxygen without any kind of haemoglobin-like molecule, relying only on oxygen dissolved into the blood.

# Humans need a solid eight hours of sleep every night

If opponents of Donald Trump's many controversial policies wonder how he sleeps at night, the answer is: he doesn't – much. 'How does somebody that's sleeping 12 and 14 hours a day compete with someone that's sleeping three or four?', he once asked a reporter. It is one of numerous boasts from the President about his ability to get by on minimal sleep. He is in illustrious company. Former British Prime Minister Margaret Thatcher could function on four hours a night. The Iron Lady would often keep officials working on policy into the small hours, and then rise with the dawn to catch the morning news. Bill Clinton was another night owl, working a similar schedule to Mrs T. Napoleon put it most pithily: when asked how many hours sleep a person needed, he is supposed to have replied: 'Six for a man, seven for a woman, eight for a fool.'

Medical science has much to uncover about sleep. We still don't fully understand why humans, and more or less any other animal you can name*, need to sleep at all. While lack of sleep can affect the immune

---

*FOOTNOTE: Most animals, including some worms and insects, require sleep – or something like it. Animals are particularly vulnerable while sleeping, so there must be some good reason why the behaviour evolved so early, and near universally.

system and hormone levels, the biggest loser is the brain. Our slumbers are thought to help strengthen and prune connections in that organ – helping to stabilize memories and purge unimportant links. Sleep is also an opportunity to flush the brain of waste chemicals that might have built up during the day. As scientists are fond of saying: more research is needed.

Lack of sleep is not much fun. We're all familiar with the symptoms: drowsiness, irritability, stress, poor concentration. As with most things in life, Keith Richards of The Rolling Stones once pushed these buttons to the limit. After nine days without any sleep, the rocker supposedly keeled over and smashed his nose. Regular periods of sleeplessness may have even more serious consequences. Studies have linked sleep deprivation to a greater chance of developing heart disease, obesity and diabetes. The immune system is also weakened, leaving the insomniac prone to infectious diseases.

But how much sleep is optimal? Eight hours per night is a common answer. That means a healthy person spends about one-third of their lifetime unconscious. When put like that, it's easy to see why workaholics and creative types want to claw back that down time. What a waste! It should be remembered, though, that eight hours is a recommendation based on an average. Everybody is different when it comes to sleep. Some people may need ten hours per night, while others will feel refreshed on six. One size of pyjamas does not fit all.

There's some evidence that a solid block of eight hours is not the natural state of affairs for humans. Our ancestors, more attuned to sunrise and sunset, would have slept for four hours, then spent a couple of hours in relaxed wakefulness, before sleeping for another four hours. References to 'first sleep' and 'second sleep' can be found all over the place in literature from before the Industrial Revolution. Everything from Chaucer's *Canterbury Tales* to doctors' notes describe the double dose as though it is a commonplace. This is probably our natural preference – limited studies have shown that subjects will revert to this pattern if kept in a darkened room for long periods. Others have suggested that the sleep gap was a practical necessity. Our ancestors – at least the ones in colder climates – had to get up in the middle of the night to keep the fire burning. Modern luxuries like electric lighting, automated heating and precise timekeeping have removed this need.

Can Donald Trump really get by on just three or four hours *every* night? Probably not. I suspect he has an occasional night of minimal sleep, followed by nights of five or six hours. Claims that he's a man who rarely sleeps might explain a lot about his erratic presidency but are more likely a mix of braggadocio and selective memory. Donald, please get in touch via my publisher if this is fake news, and you'd like to put the record straight.

# Women have more ribs than men

Men and women have precisely the same number of ribs: 24, in 12 pairs. A persistent myth would have us believe that women have two more than men. The misconception is usually attributed to the Book of Genesis, which describes the first and most impressive example of tissue engineering:

*And the Lord God caused a deep sleep to fall upon Adam, and he slept: and he took one of his ribs, and closed up the flesh instead thereof; And the rib, which the LORD God had taken from man, made he a woman, and brought her unto the man.* (Genesis 2:21–22)

This passage doesn't really imply that Eve ended up with more ribs, only that Adam lost one in the creation of Eve. In any case, their children would not be affected by the supernatural surgery. Adam's DNA would have furnished his offspring with a full complement of ribs, no matter how many bones the Lord God teased from his body.

It is possible to have more than the standard 24 ribs. 0.5–1 per cent of people have cervical ribs – additional bones that develop at the top of the rib cage, from the neck. These are usually harmless but can pinch nerves in some cases. Curiously, cervical ribs may be more common in women than men (J. Brewin et al., 2009, doi:10.1002/ca.20774). A determined pedant could therefore make a case that women really do have more ribs than men, if only fractionally, and when averaged across a large population.

# Same thing, different name

Medics often use fancy-pants terminology to describe body parts. They deal in patellas and pollices, where the rest of us see kneecaps and thumbs. They examine clavicles when consulted about collar bones. They know their intergluteal cleft from their olecranon.

They are not being snooty or needlessly grandiloquent. This language, often derived from Latin, has two advantages. It can be more precise: 'mandible' is the medical name for 'jaw', but mandible distinguishes the lower jaw specifically. The other reason is ease of communication with colleagues from different parts of the world, where everyday names for a certain body part might differ. It's not 'wrong' to call a patella a kneecap, but the medical term can have a greater degree of accuracy. Here are just a few.

| Common name | Medical name |
| --- | --- |
| Armpit | Axilla |
| Belly button/navel | Umbilicus |
| Big toe | Hallux |
| Breastbone | Sternum |
| Collarbone | Clavicle |
| Elbow | Olecranon (the bony, pointy bit) |
| Jaw (lower) | Mandible |
| Kneecap | Patella |
| Kneepit (back of knee) | Popliteal fossa |
| Little toe | Minimus |
| Mole/birthmark | Nevus |
| Shoulder blade/wing bone | Scapula |
| Tailbone | Coccyx |
| Thumb | Pollex |

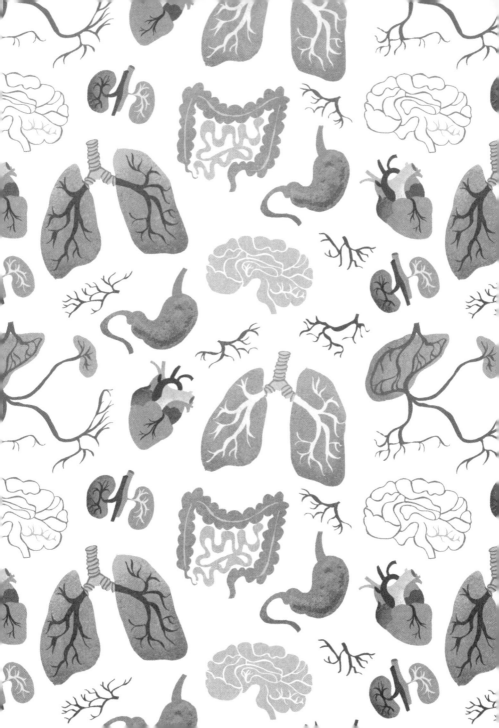

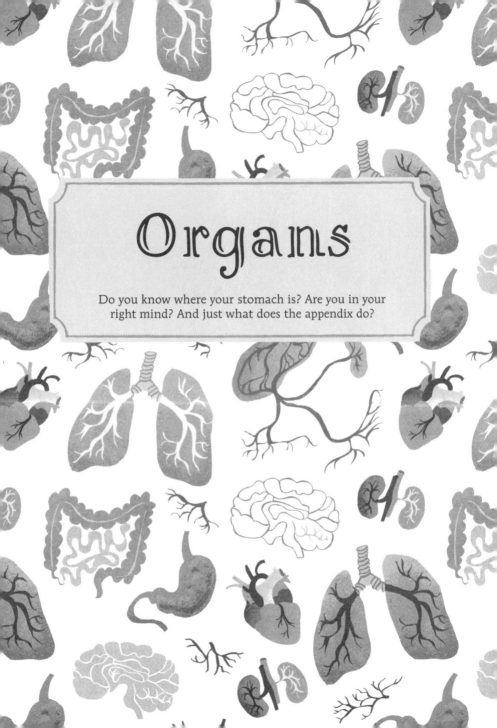

# Organs

Do you know where your stomach is? Are you in your right mind? And just what does the appendix do?

# Your stomach is down behind your belly button

Our innards are mysterious. Most of us never get to see our own stomach, liver, kidneys, brain or bowels. It's a peculiar fact when you think about it. The vast majority of your physical being is never glimpsed by anyone. This inner ignorance has amusing side-effects. How often have you heard someone say they have a baby in their tummy (an abbreviation for stomach)? Just think about that. The stomach, with its highly acidic, digestive environment, is the very last place you'd want to stow a foetus.

It's common to think that the stomach lurks somewhere behind the belly button. In fact, it sits like a champion on top of the rest of the gut. It's much higher in the abdomen than most of us realize. Find the point where your lower ribs meet in the middle, then go three finger widths down and three to your left. You should be right over the centre of your stomach. It's closer to the nipples than the belly button*.

When we say we have stomach ache the pain we feel may actually be from some other part of our gut. You won't be surprised to learn that the sensation of 'butterflies in the stomach' doesn't involve actual

---

*FOOTNOTE: As a child, I recall feelings of terror that my belly button might come undone. The idea that the knotty naval nodule might untangle, with gory consequences, remains common, and not just among children. Fortunately, the belly button is no more likely to open than any other point on your skin. It's not a knot, but scar tissue left over from the fused stump of the umbilical cord.

butterflies. More unexpectedly, it has little to do with the stomach either. The fluttery feelings stem from the intestines. During a tense situation, blood supply to the intestine muscles is diverted to places better suited to escape, like the leg muscles. It's the famous 'fight or flight' response. The stomach will be affected by this internal shift too, but it is mostly an intestinal phenomenon.

The sensation of 'heartburn', meanwhile, usually has nothing to do with the heart. It's caused by stomach acid sneaking up into the food pipe (oesophagus) where it does not belong. The oesophagus lacks a protective lining, and so the acid can cause minor damage and therefore pain. We call this 'heartburn' because the junction of the stomach and oesophagus is close to the heart. So close, in fact, that the two systems share the same nerve supply. Hence, the symptoms of heartburn can feel similar to those of a heart attack, even though they have very different underlying causes.

The locations of other organs can be surprising. The heart is often assumed to be way over to the left of the chest. It's better to think of it as centrally placed, but bulkier on the left. Were Welsh songstress Bonnie Tyler to surgically demonstrate her biggest hit, 'Total Eclipse of the Heart', then the shadow would fall partly to the right of centre as well as the left.

The liver, too, is higher in the body than often imagined. It resides just beneath the right lung alongside the stomach. The kidneys are up there too, nestling beside the liver and spleen. Because of the way the nerves are wired up, kidney pain is often felt lower down, closer to the bladder or, in men, even the testicles.

Some people have organs where even their doctors don't expect. In a rare condition called situs inversus, the organs are mirrored from their normal positions. The liver, for example, is over to the left, while the spleen and stomach march right. Situs inversus usually has no side effects. Those with the condition can be totally unaware that they possess transposed bodies until they encounter a medical problem – such as a burst appendix on the left rather than the right. Organ transplants can also be complicated, thanks to the inverted plumbing.

# The liver is the largest organ

The typical adult liver weighs about 1.5kg (3.3lb). It's one of the largest structures in the body – about the same shape and size as an eggplant. The liver is the body's jack-of-all-trades. It has something like 500 functions. These include the breakdown of toxins, storage of energy reserves and production of bile. Best not look now, but you'll find it above the stomach and below the lungs – where your rib cage begins to peter out.

The liver is the largest internal organ, heavier even than the brain. It is not the largest of all organs, however. That accolade belongs to the skin. This might not seem like an organ. It's not a blob you can hold in your hand,

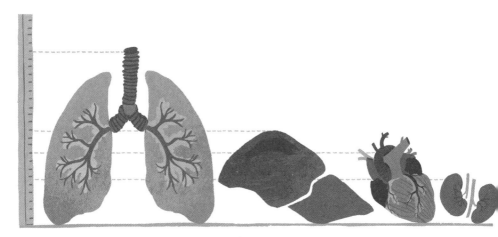

or order from your butcher like a liver, heart or kidney. But the skin is a self-contained part of the body with specific functions, so it is classified as an organ.

Were you to flay an adult human, the excised skin would cover about 2sq. m (21½sq. ft). That's enough to horrifically conceal half a ping-pong table*. In terms of mass, the skin makes up about 16 per cent of body weight. The absolute weight depends, of course, on how big you are. A typical value is around 4.5kg (10lb) – as much as a small microwave oven. If that seems surprising, consider that your skin varies in thickness around your body. Areas such as the underside of the foot and the base of the thumb are particularly chunky. It all adds up to make the skin the heaviest organ by some margin.

Some unfortunate individuals have still-heavier body parts. While it would be a stretch to call them organs, tumours can grow to physically debilitating size, far eclipsing the liver or skin in weight. The largest on record was removed from the abdomen of an unnamed lady in California in 1991. It measured a full metre (3ft) in diameter and weighed a colossal 136kg (300lb). Following surgery, the patient weighed 95kg (210lb), significantly less than the growth.

---

*FOOTNOTE: As far as I'm aware, human skin has never been put to this purpose. It has, however, found many other grim uses over the centuries. In the 19th century, hanged criminals were occasionally flayed for book leather. The notorious murderer and grave-robber William Burke was perhaps the most famous, but several other cases are known. Human skin has also been adapted for shoes, wallets and even a drum head.

# Humans only use 10 per cent of their brains

The human brain is unique, so far as we know, in the whole Universe. No other object is capable of such breathtaking versatility. What other structure can control knitting needles, imagine faces that never existed, fall in love or appreciate jazz – all at the same time?

How the vast collection of cells within your skull somehow club together to achieve consciousness, emotion, rational thought and self-awareness can feel tantamount to magic. With mystery comes credulity. You could make up almost any nonsense you like about the brain and find a nodding audience. This is probably why the factoid that we only use 10 per cent of our brains has such traction. The suspect statistic also suggests that 90 per cent of the brain lies dormant, just waiting to be harnessed. Cue a deluge of self-improvement books.

It's a silly fact, though, even at face value. I mean, what is meant by 'use'? I could say that I only use 5 per cent of any given tooth – the surface that does the biting or munching. Does that mean that the rest of the tooth is useless? No. It plays a structural role, holding the cutting edge in place. Teeth also help to shape speech. A flash of the teeth is a key part of body language to indicate anger or fear. It's not just the tip that's 'used'. We need to consider the tooth, the whole tooth, and nothing but the tooth.

And so it is in the brain. We've all heard of neurons – the brain cells that form complex connections, allowing control, response, memory and thought. Surprisingly, these only make up about half the brain\*. Glial

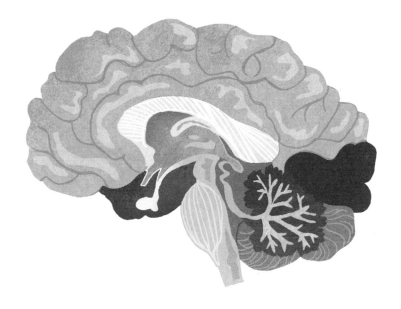

cells get much less attention, but we each have 100 billion of them. These cells do not carry nerve impulses but play support roles. Some provide the neurons with nutrients and oxygen; others serve as cleaners, removing toxins and dead cells. Still others act like scaffolding and hold neurons in a particular orientation. Others wrap around the neuron to form an insulating sheath. In this sense, you don't use half your brain's cells for

---

*FOOTNOTE: The proportion of glial cells to neurons was long thought to favour glial cells by as much as 10:1. More recent research (see C.S. von Bartheld et al., 2016, doi:10.1002/cne.24040) suggests it is closer to 1:1.

the kind of tasks usually associated with that organ, but that is a different thing to saying half the brain is unused.

The myth is often stated as 'We only use 10 per cent of our brain power', which suggests that we're talking about neurons. If we consider only neurons, then, does the story ring true? Not really. There is still a great deal to learn about brain function, but it seems clear from scans that all regions of the brain are in use during some part of the day.

That doesn't necessarily imply that all the neurons within those regions are firing. A football crowd of thousands will still sound pretty loud if only 100 of them are chanting. It's (currently) impossible to monitor every single neuron in the brain (after all, there are billions of them), so we can't know exactly how many are active at any given time. By the same token, there is no way to show that exactly 10 per cent are active either. Given that the myth dates back to at least the early 20th century – a time without any form of brain scan – the number is clearly plucked from thin air with no basis in measurement whatsoever.

Even so, it is possible to live happily without certain parts of the brain. Surgery to remove whole regions, such as a temporal lobe, is sometimes necessary for the treatment of seizures. This comes with wretched effects, including memory loss and language problems. The brain, though, is a very adaptable organ. It rewires to compensate. Some patients who undergo this type of brain surgery even report an increase in IQ.

In an extreme procedure known as a hemispherectomy, surgeons remove or disable half of the brain. Let me say that again in case you were skimming: a whole hemisphere of the brain can be removed without killing the patient.

This last-resort surgery is only helpful for those who have daily seizures in the same half of the brain. It works well. One review found that 86 per cent of children who undergo the procedure no longer had seizures. Remarkably, they suffered few long-term cognitive problems, although many lost vision in one eye and movement in one hand. Some of the children even showed improvements in the classroom – presumably thanks to the sudden absence of seizures.

The last two paragraphs would seem to support the notion that we only use a portion of our brains, even if it is more than 10 per cent. But saying that we can live without large chunks of the brain is not the same as saying we don't use those chunks. I could still drive my car if a peculiar thief removed my passenger seats, windows, air con and radio. That doesn't mean I have no use for those parts.

Still, the myth is readily trotted out all over the place. It's quite a handy device for certain individuals. Those selling self-help books can appeal to this 90 per cent as a reservoir of untapped potential, which only they can teach you to syphon. Mystics who believe they have psychic powers (or want you to believe they have psychic powers) can also invoke the unused brain canard. We might all be able to read minds, speak to the dead, see into the future or bend spoons, if only we could learn to use that other 90 per cent.

Oddly, those who claim to have access to these dormant brain resources are rarely, if ever, among the world's top scientists, philosophers, mathematicians or any other kind of intellectual profession. If they're actually able to tap into additional intellect – and perhaps even read minds – how come their CVs never stretch far beyond motivational speaking or psychic sideshows?

# You're either a left-brain person or a right-brain person

Have you ever been told that you're a right-brained person? You probably read this in a magazine or on a Facebook app after answering a series of multiple-choice questions about yourself. If you're right-brained you tend to be a bit arty; you enjoy being creative, socializing and singing out loud at the bus stop. If, on the other hand, you're a more logical thinker, making lists and solving puzzles, then you are a left-brained person. You definitely don't sing at the bus stop. You probably analyze the timetable and raise an eyebrow at the optimistic journey times between destinations.

It's a common understanding that our personality can be split in twain: arty skills are controlled by the right hemisphere of the brain, and more analytical skills come from the left. You might use your right brain to create an impressionistic painting of a wheat field, while your left-brained friend makes a spreadsheet of crop yields.

It's nothing like as simple as that. The human brain is too complex, too interlinked to be so crudely generalized. The idea that Michelangelo made more use of his right brain, while Newton had an overdeveloped left brain is nonsense. Such broad statements are a staple of pop-psychology magazines but have no grounding in reality.

It's true that certain mental tasks are performed predominantly in one area of the brain. For example, most aspects of speech are controlled from the left hemisphere, while the right side handles much of face recognition. But your personality type is not determined by a particular hemisphere. Imaging experiments have shown that individuals do not typically favour one side of the brain over the other. On most tasks, the two hemispheres work together.

This makes sense when you think about it. Imagine again that you're painting that wheat field. Your brain needs to exercise precise control over your arm and hand muscles. You must calculate the best angle and force with which to paint the canvas, select the appropriate brush, and choose the hour of the day that offers optimal lighting. Your imagination will play upon the wheat field, conjuring up colours, shapes and impressions that don't map directly onto the witnessed scene. Your brain must translate a three-dimensional reality onto a two-dimensional surface. You must mix the paints on your palette. Which of these tasks involve the left brain and which the right?

You probably also have a motivation for making the painting in the first place, a sense of the tradition of painting outdoors, and an appreciation

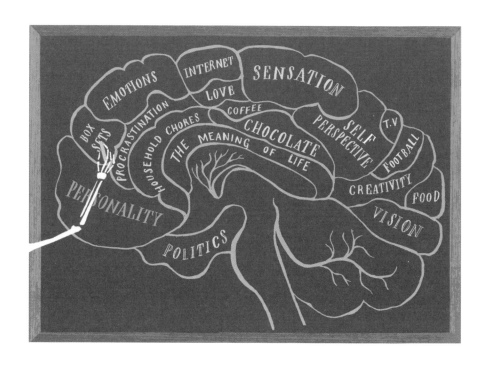

of some of its greatest practitioners, who you would like to emulate. You weighed up the cost of your training and materials. You told a lie to your boss in order to find the time to come out here and paint. At the end of the day, you fire up Instagram to show off your creation to the wider world. A painting is not simply an act of creativity; it demands many skills from its maker, some of which are sparked from the left brain, some from the right, but most from an interplay of both.

Incidentally, the old anecdote that left-handed people are more creative than right-handed people is also suspect. Yes, one can readily find lists of geniuses who favoured that hand: Leonardo da Vinci, Wolfgang Mozart, Marie Curie, Bill Gates and Albert Einstein, for five. Leaving aside that da Vinci and Einstein were probably ambidextrous, you can easily counter with a list of right-handed talent. To draw from similar fields, we might offer Titian, Gustav Mahler, Louis Pasteur (who, incidentally, noticed that chemicals could be left- or right-handed), Elon Musk and Niels Bohr.

It's difficult to prove that left-handedness is linked to either intelligence or creativity. These qualities are hard to define and measure. However, some of the statistics are intriguing. Although only 10 per cent of the population is left-handed, 25 per cent of Apollo astronauts favoured that hand. Four of the last six US Presidents (Reagan, G.H.W. Bush, Clinton and Obama) have been left-handed or ambidextrous.

# A flatlining heart can be restarted with a defibrillator

Prepare to be shocked. A flatlining heart cannot be restarted with an electric charge.

This goes against almost every medical drama you've ever seen. It also confounds common sense. Surely a heart that's lost its electrical activity should be zapped back to life with a defibrillator? This would be analogous to jump-starting a car by using an external source of electricity to bypass a flat battery. It would also be wrong.

The electric pulses that regulate heart rhythm rely on the coming and going of charged particles – chiefly calcium, sodium and potassium ions. These elements pass into and out of heart cells through dedicated channels in the cell membrane. As they do so, the voltage across the membrane changes, causing the cell to contract or relax. Neighbouring cells are encouraged to join in. A wave propagates through the heart as successive cells contract and then relax. The heart is beating; the blood is circulating.

The ebb and flow of electrolytes in heart cells is rather beautiful, and worth studying further if you get the time (try an online video – it is hard to keep track of the various ions from a written description). The dance never ceases. Until it does, and then you are in dire medical trouble.

When somebody flatlines, it means that no electrical activity can be found in the heart. The complex dance of the electrolytes has stopped or dwindled to undetectable levels. Applying an electric shock can do nothing to help. It's like trying to jump-start a car that doesn't have any fuel.

Occasionally, flatlining patients can be resuscitated. Medics will try to pump the heart from the outside by pressing down repeatedly on the chest (CPR). The patient will be given adrenaline or something similar to encourage a rise in blood pressure. Meanwhile, others will try to figure out why the heart stopped beating in the first place – a blocked blood vessel (embolism) or very low blood pressure for example. At no point will anyone shout 'Clear!' and apply the defibrillator paddles.

Electric shocks only work if the heart has some kind of electrical activity – a 'shockable rhythm'. The most common scenario is when the heart stops its familiar 'lub-dub' and instead goes into a quiver. The individual cells still pump electrolytes, but not in a coordinated fashion – like a sports crowd whose members are all chanting different songs. Shocking a heart with a defibrillator forces all the cells to contract at once, resetting them to the same state. If successful, the heart resumes its regular rhythm.

A trained medic will only wield the defibrillators if the heart monitor shows irregular activity. A TV medic is looking only for a flatline. It is easier to find dramatic tension in a sudden stop than in a stuttering dwindle. Viewers expect to see the complete loss of beat followed by that familiar tone of death. This is a myth made entirely in the film studio.

# The appendix has no known function

The poor old appendix has a reputation as a useless organ. The worm-like feature extrudes from the large intestine; a tube that goes nowhere. Even its name, shared with the part of a book that holds surplus information, suggests the inessential. Its only role is to make otherwise healthy people scream with pain when it decides to go wrong. At best, the appendix is a 'vestigial organ' – a bit of digestive kit that was useful to our hunter-gatherer ancestors but has since become redundant and withered by changes in human diet. It can – and often is – removed without any ill effects. This is a pointless organ.

Or so people thought. The appendix, it now seems, is a reservoir for helpful bacteria. Microbes that help us fight disease are stored there,

ready to be deployed if an illness wipes out their brethren in the wider gut. This has been tested clinically (G.Y. Im et al., 2011, doi:org/10.1016/ j.cgh.2011.06.006). Patients without an appendix were significantly more likely to have a recurrence of the gut infection caused by the bug *Clostridium difficile* than those with an appendix. The conclusion drawn was that it helps restore the gut's population of beneficial bacteria after the first infection, helping ward off any future attacks. It should be noted, however, that the connection is still not definitive.

The appendix also plays a part in the body's immune system. During our early years, the appendix helps out with the formation of white blood cells and certain types of antibody. In the fetus, the appendix produces peptide hormones that feed into homeostasis, the complex series of interactions that keep the body's internal environment in balance. These roles diminish as we get older.

Those who have their appendix removed usually go on to live a completely unimpeded life. The appendix, though, is still worth keeping if at all possible. The organ is sometimes used by surgeons as a spare part. It can be fashioned, for example, into a ureter – the tube that carries urine from the kidneys to the bladder.

The appendix is not the only part of the body we can live without. Teeth, fingers, hands and limbs are clearly not essential to life, but many internal organs are also expendable. Humans can function perfectly well with only one kidney. Likewise, a whole lung can be removed with only minor consequences. You only get one spleen, but it's very much a 'nice to have' rather than an essential. Those without one usually lead normal lives, but with an increased risk of infection. The stomach is also non-essential – with the right surgery, the intestines can cope with most foods. All reproductive organs can be jettisoned without ill effect, other than the obvious. Curiously, you can remove more than one-quarter of the skeleton and still lead a full life – those without feet are missing 52 of their 206 bones. You can even get by with half your brain removed, as we saw earlier in this section. Quite a lot of what we carry around isn't strictly necessary.

# A miscellanea of dodgy folk tales

This section looks at what were traditionally known as 'old wives' tales'. The phrase is somewhat dated. It seems to me that men peddle these anecdotes as often as women, and age has nothing to do with it.

**Bald men are more virile:** Bruce Willis, Samuel L. Jackson, half the cast of *The Fast and the Furious* films ... there's a certain masculine, macho image that goes alongside the bald pate. A hairless scalp has long been linked to high levels of testosterone, the primary male sex hormone. There is indeed a limited connection. Men who produce no testosterone (for example, those who have been castrated) do not go bald. Those who do produce the hormone (the vast majority) can go bald, but incidence seems to be independent of testosterone levels. Even a small amount can do it. Whether a man will go bald is down to other factors, which have nothing to do with virility, sex drive or ability to race cars at high speed.

**Carrots help you see in the dark:** Of all the tall tales in this book, the idea that carrots aid night vision has the most unusual provenance. The story was supposedly made up by Britain's Air Ministry during the Second World War to account for the Royal Air Force's uncanny ability to find enemy bombers at night. In reality, British pilots were using a new form of radar to spot approaching squadrons. The government didn't want the Germans to know about this top-secret technology, and so word was put out that British pilots ate large numbers of carrots to help them see in the dark and shoot down more bombers. It's a great story, but hard to verify. The government certainly did encourage both pilots and the general public to eat carrots to help with night vision. But was this really a deliberate attempt to distract the Germans away from radar? That rumour started

soon after the war but is never backed up with evidence – even on debunking websites like Snopes.com.

But back to the carrots themselves. Can they really boost night vision? The dietary advice is sound up to a point. Carrots contain high levels of vitamin A or retinol, a compound that helps eyes adapt to dark conditions. Eating the vegetables will ensure normal night vision. Anybody living under blackout or, indeed, piloting an aircraft at night, would be well advised to eat carrots and other vegetables to keep their night vision healthy. But carrots cannot improve vision beyond that norm. The idea is still widely believed and, in some ways, rather useful. How many children have been persuaded to eat carrots in the belief they'll get superpowers?

**Eating cheese gives you nightmares:** A particularly memorable line from Charles Dickens's *A Christmas Carol* has Scrooge blaming his ghostly visions on his supper: 'You may be an undigested bit of beef,' he tells the ghost of his former business partner, 'a blot of mustard, a crumb of cheese, a fragment of underdone potato. There's more of gravy than of grave about you, whatever you are!' That's not only the worst pun in the whole of literature, it's also a myth. There is no evidence linking cheese, or any other foodstuff, to the quality or quantity of our dreams. It may be that eating late affects sleep patterns, as the body works to digest the food, but no correlation to nightmares or any other kind of dream has ever been proven.

**Swallowing bubble gum is dangerous:** This myth comes in as many flavours as the product itself. Some say that gum can't be digested and hangs around in the gut for seven years. Others reckon it will stick to your insides and cause a blockage, or else tangle up the intestines. Sorry to burst the bubble, but there are few health risks to swallowing gum. It's true that the body can't digest most of it, but that's no issue so long as you don't swallow packets of the stuff in one go. The undigested gum will simply catch a ride into your toilet bowl within a couple of days of being swallowed.

**If you get a nosebleed, you should tip your head back:** This measure might limit the amount of blood that drips down onto the floor, but it will not have much effect on the malady itself. If you tilt your head backwards, the blood will instead trickle down your throat. That presents the risk (a small one, admittedly) that you might black out and choke on your own fluids. Better to sit it out with a wad of tissues pressed against the nostril.

**Masturbation causes blindness:** The myth usually refers to male masturbation. Further possible symptoms include impotence, infertility, baldness, deformity and – bafflingly – hairy palms. There are no risks to physical health.

**You can get pregnant from toilet seats:** There is no known case of anyone ever getting pregnant from a toilet seat. It is, I suppose, technically possible. You'd have to (a) be female, (b) sit on a seat onto which somebody had very recently masturbated, (c) find some way to collect the ejaculate, (d) perform some form of self-insemination, (e) be very fertile. The full sequence is not very likely unless you have some unusual toilet habits.

**It takes more muscles to frown than to smile:** This old aphorism is a pseudo-scientific way of saying 'cheer up'. Sometimes the number of muscles is given: 50 to frown, 13 to smile, for example. These numbers change with every telling, a sure sign that something is a bit iffy. The myth seems to have started in the 1920s, when it was commonly used in newspaper adverts. A non-sequitur logic would have you buy a dress (or a photography session, or some toffee – it really was an over-exploited phrase) because it would make you smile, and scientists say that smiling uses fewer muscles. In those days, the ratio was consistently 62 muscles to frown and 15 to smile, but is now any random pair of numbers. The true answer is impossible to determine. It depends what constitutes a smile and a frown, both of which take many forms. Plus, even if a smile does use fewer muscles, they may be ones that require greater exertion to flex.

**Waking sleepwalkers can be harmful:** The risk here is that the startled somnambulist might suffer a heart attack from the sudden shock. There is little danger of this – no more than from waking up to an alarm clock. It's far more likely that the person will hurt themselves while fumbling around in the dark. Ideally, the oblivious sleepwalker should be guided back to bed, but waking them will do no harm.

**You can catch warts from toads:** Most city dwellers rarely see a toad, let alone worry about the risk of contagion. But lest there be any doubt, there is no way to catch a wart from a toad. The wartish lumps commonly seen on the amphibian's back are actually glands. They secrete toxic chemicals as a defense mechanism. Human warts are, by contrast, caused by the human papillomavirus. You can only catch it from other humans. Not toads.

**You know what they say about men with big feet?** The idea that a man with large feet will also possess a substantial penis is hearsay. Somebody actually did the science. A 2002 study (Shah and Christopher, 2002, doi:10.1046/j.1464-410X.2002.02974.x) at University College London found no correlation. For any male readers who want to check their own non-vital statistics, the average 'stretched penile length' is 13cm (5in), while the average UK shoe size was 9 (European 43, US 11).

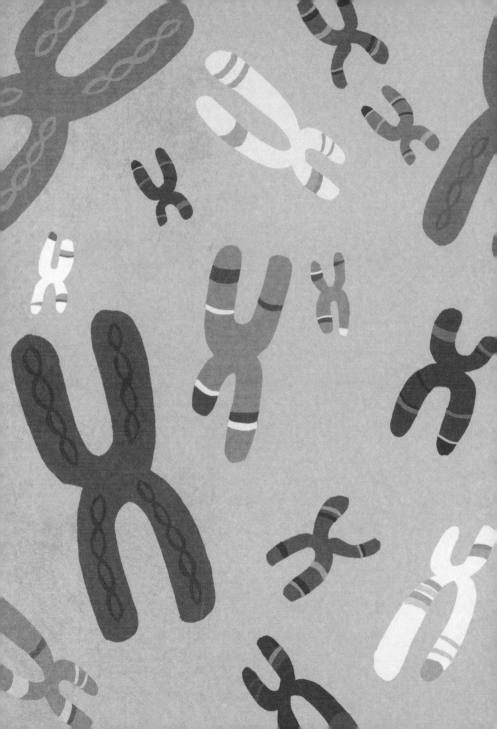

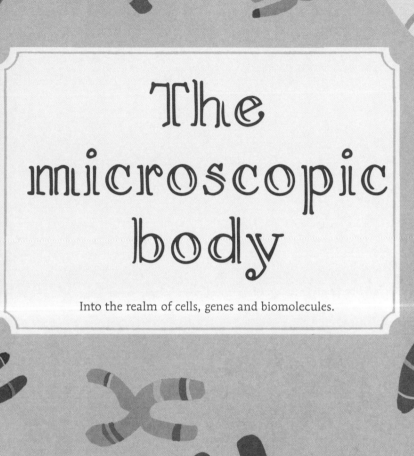

# The microscopic body

### Into the realm of cells, genes and biomolecules.

# All the cells in your body contain 46 chromosomes

DNA in our cells doesn't just float around willy-nilly. It is tightly and efficiently packaged. If you peered at DNA under a bench microscope, you wouldn't see the famous double helix. The genetic material is coiled up around proteins known as histones, like thread around spools. These then fold into yet more complex patterns until eventually they form a chromosome. Chromosomes are often drawn as elongated cross shapes. Down a microscope, they tend to look like undifferentiated blobs – only taking on the appearance of a cross during cell division. Most human cells contain 23 pairs of chromosomes, or a total of 46. The vast majority of cells, in the vast majority of people, conform to that tally. However, there are many exceptions.

The most common, of course, are the sex cells. Sperm and eggs only contain one copy of each chromosome. They have 23 chromosomes in total and are said to be 'haploid' – as

opposed to other cells, which are 'diploid'. When they fuse at fertilization, the two sets come together to make a 'diploid' cell – the new ball of life has a full complement of 46 chromosomes.

Certain genetic conditions can also alter the number of chromosomes in the cell. Down syndrome also goes by the name of trisomy 21, which offers a clue to its genetic origins. Those with the syndrome carry an extra copy of chromosome 21 in every diploid cell – they have 47 chromosomes. An extra chromosome might not sound too traumatic, but its presence disturbs the delicate balance of protein synthesis. That has knock-on effects all over the body. Individuals with Down syndrome are prone to numerous health issues, as well as characteristic physical differences.

An additional X chromosome* in men – again giving 47 chromosomes in total – leads to a condition known as Klinefelter syndrome or XXY. The consequences are usually less marked than with Down syndrome. Men with the additional X tend to have smaller testicles and may be infertile, but many individuals are unaware of their genetic distinction. In rarer cases, the man may have more than one extra X chromosome, with more severe symptoms. Likewise, about one in a thousand men are born with

---

*FOOTNOTE: You've probably heard of the X and Y chromosomes. These were traditionally seen as the chromosomes that determine sex. Someone with two XX chromosomes is genetically female, while an XY carrier is genetically male. We now know that this simple binary system is only part of the story. Sex determination depends on the complex interplay of many factors, not all of which are dictated by the X and Y chromosomes. Indeed, it's possible to carry X and Y chromosomes and have female genitalia, or two X chromosomes and appear male. In addition, not every gene on the X and Y chromosomes is linked to sexual characteristics. For these reasons, scientists today tend to avoid the phrase 'sex chromosomes'. Also contrary to common assumption, the Y chromosome is not 'Y'-shaped. Its name was chosen purely as an alphabetical counterpart to the already labelled X chromosome with which it pairs.

an additional Y chromosome. Again, the repercussions are usually not severe. Men with the condition tend to be taller than average, with an increased risk of learning difficulties.

Some people have fewer than the usual 46 chromosomes. Turner syndrome, which affects females, results from a missing or partially missing copy of the X chromosome. The condition comes with a raft of signs and symptoms, with physical traits that may include a wide neck, low-set ears and short stature. Females can also carry additional copies of the X chromosome. Triple X syndrome is reasonably common, affecting around 1 in 1,000 female births, and usually without serious implication. Less common, and often more debilitating, are XXXX and XXXXX syndromes. In the latter case, individuals have 49 chromosomes in each cell.

Incidentally, our default tally of 46 chromosomes is not particularly bountiful when compared to other species. A goat has 60 chromosomes; a dog 78. Even cigarettes are better endowed. The tobacco plant boasts 48 chromosomes, pipping us by two. Certain lampreys (eel-like fish) have 174 chromosomes. Ferns can blast past the 1,000-mark.

# All the DNA in your body is found in your chromosomes

Every cell in your body contains a copy of your DNA – the 'instruction manual' or 'blueprint' for putting together the proteins of your body. You're no doubt familiar with its underlying shape – the famous double helix. This in turn is coiled up tightly around proteins known as histones. And, as we've seen, this tangle of tangles is then folded still further into structures known as chromosomes.

Everybody's DNA is unique and set at conception by the intermingling of DNA from the mother and father. Of the 23 pairs of chromosomes in a human cell nucleus, one of each pair comes from the father, and one from the mother.

Not all of your DNA is wrapped up in chromosomes, though. It doesn't all sit within the cell nucleus. Nor does it all come from your mother and father. Much of the DNA in your body isn't human at all. If we could mulch down a human being to a soup of biomolecules, and then assay the mess for different types of DNA, we would find a sub-microscopic menagerie of thousands of species. Here is a probably incomplete catalogue of the more unusual varieties of DNA in your body.

**Mitochondrial DNA:** Most human cells contain mitochondria. These small organelles are often known as the powerhouse of the cell. It is here that ATP molecules – the cell's chief source of chemical energy – are produced. The tally varies from cell to cell. Red blood cells are entirely devoid of mitochondria, while liver and muscle cells can have thousands. Mitochondria are peculiar structures. They were once free-living beings, similar to bacteria. Then, around 1.5 billion years ago, they somehow fused with a larger cell, creating a symbiotic partnership that still underlies our cells to this day. Mitochondria contain their own DNA, separate to the chromosomal DNA of the nucleus. While nuclear DNA arranges into 'X'-shaped chromosomes, that in the mitochondria is in the form of a hoop. This ringlet of DNA is tiny compared with the regular chromosomes. It contains just 16,569 base pairs*, whereas even the smallest chromosome (the Y chromosome) has more than 57 million. The lowly mitochondrial DNA makes up something like 0.0005 per cent of your genome. Even so, the proteins it codes for are critical to the cell's energy supply. Strip away these minikin hoops, and you will die instantly.

**Microchimerism:** Your body contains something like 37 trillion human cells. The vast majority are truly yours. They harbour the same 46 chromosomes and the same mitochondrial DNA as the first fertilized cells from which you ripened. You're not the only one in there, though. Your body contains the cells of other humans, and not just the temporary residues from those you get particularly close to. Welcome to the strange and little-understood world of microchimerism.

We can all fondle our belly buttons as a reminder of our first nine months in the womb. Mothers, too, carry an internal remnant of each child they have borne. During pregnancy, cells from the baby find their way into the mother. They don't hang around near the umbilical cord, but migrate to

---

*FOOTNOTE: Base pairs are the main building blocks of DNA. If you're unfamiliar with this terminology, then a common analogy is to think of them as the letters that make up a book.

different corners of her body, where they quickly adapt to match the local tissue type. For example, cells might reach the heart, latch on, and turn into cardiac tissue. The intruder cells are genetically distinct from the mother's cells, but can remain in situ for decades, functioning and dividing alongside the mother's native cells.

This raises some intriguing possibilities. A mother may take in cells from multiple pregnancies, even ones that don't go to term. A child lost at birth may, in a very real sense, remain in the heart forever, if some of its cells migrated to that organ in the mother. The process works two ways. Less frequently, cells from the mother pass into the foetus. This being the case, the baby might also take in cells from its mother's previous pregnancies. If you have an elder brother or sister, it's quite possible that you carry small amounts of their DNA. As the cells can persist for decades, they might even be passed on to a further generation. A piece of your uncle may linger in your shoulder; a tincture of grandma may sequester in your kidney.

Microchimerism is a relatively recent discovery and is still poorly understood. We do not know if it has a role in disease, immunity or has no consequence at all. But one thing is clear: at the cellular level, we are not entirely ourselves.

**Microbial DNA:** The 46 chromosomes and mitochondrial hoops make up all the DNA in your genome, and we've seen that DNA from close relatives can also tag along. So far, so human. But were we to fish out all the DNA in the body, we would find plenty of genetic material from other species. Thousands of other species. We think of ourselves as individuals, but really each one of us is a slew, an aggregate, a collection of countless beings. Your every flange, flap and crenellation teems with microscopic life. This is the 'microbiome', and it includes bacteria, archaea, viruses and fungi.

Your outer surfaces support an enduring ruckus of bacteria, no matter how exemplary your toilet habits. Your skin is crawling with them. The belly button alone supports an average of 67 species of bacteria. Bacteria also thrive within the body, particularly the gut. The ones that cause disease are vastly outnumbered by those with resident status. Some 500 different species call your colon home. Many perform useful roles, aiding digestion, bolstering the immune system, and neutralizing toxins.

Scientists now estimate that around half the cells in the body are microbial*. That is a freaky thought. Half of your body is not your body at all. Bacterial cells, however, are much smaller than most human cells. If we instead measure by mass (not including water), you are roughly 99.3 per cent human, and 0.7 per cent bacterial. Scientists are only just beginning to piece together just how important the microbiome is for health. The answer seems to be 'very'. It may turn out that genes carried in bacteria have more influence than our own genomes in determining whether and how quickly we regain health during illness.

Archaea look superficially like bacteria, but they are utterly different at the molecular level. A hippopotamus and a daisy have more in common than a bacterium and archaon. We harbour fewer archaea than bacteria, but they do play important roles. One, called *Methanobrevibacter smithii*, helps break down complex sugars in the gut, for example.

---

*FOOTNOTE: The figure was, until 2016, thought to be more like 90 per cent – a number still widely quoted in magazines and websites. A more detailed study (R. Sender et al., 2016, doi: 10.1371/journal. pbio.1002533) concluded that only half the cells in a human are not human. The proportions remain educated guesswork, but any of the published ratios are eyebrow-raising.

Fungi, such as yeasts, are also present in the gut. We also find them on the surface of the body. The most celebrated, if that's the right word, is athlete's foot – an itchy accumulation most commonly found between the toes. Something like 15 per cent of the world population have mould-infested feet. Fungus can grow in other places too. Ringworm (not actually a worm), jock itch, vaginal thrush (not actually a thrush), certain types of nappy (diaper) rash and dandruff all involve fungal foul play. A recent 'fungal census' found 80 different genera on a typical, healthy human body (K. Findley, et al., 2013, doi:10.1038/nature12171).

Viruses are much smaller than bacteria – little more than a short piece of DNA (or a single-stranded variation called RNA), wrapped in protein. As such, they can exist in even greater numbers within the human body. Whenever we are infected with a virus, such as influenza, trillions of copies are produced each day. Even a healthy human body will typically support five types of virus, each present in billions of copies. The bacteria in your body are themselves prone to viral infection, increasing the number to heights even more ungraspable. Viral DNA has even found its way into our own blueprints. An estimated 8 per cent of the human genome has its origins in the viral world. These genetic sequences were integrated many millions of years ago and have long since mutated into harmless passengers.

**Little creatures:** Your face is crawling with animals. Tiny mites called demodex thrive in our skin pores, follicles and sebaceous glands. They look a bit like severed fingers, with eight legs. It seems that every human has them. Demodex were first spotted as far back as 1842 but remain mysterious. It is not certain what they eat. Nor have they been strongly linked to any disease or skin problem. Nobody knows for sure how many of these hitchhikers we carry. For present purposes, it's fair to say that demodex make a tiny but unusual contribution to the DNA we carry about.

We all have demodex, but other tiny animals can take up residence in our bodies. Head lice, body lice and pubic lice are the commonest parasites in the developed world – most of us will be affected at some point. The inside of the body is also prone to parasitic infection. About half of all humans are unwitting carriers of parasitic worms*. Most cases are in developing countries, with poor access to fresh water and sanitation. Even so, about 20 per cent of people in richer nations will encounter the pinworm at some point. These charming creatures are usually discovered when we follow up a tickly, crawling sensation around the anus, where they like to lay their eggs.

So, what's the conclusion? Any human body is so much more than a human body. Thousands of microscopic species, and even tiny animals share your journey through life. When Neil Armstrong and Buzz Aldrin landed on the Moon in 1969 they were accompanied by hundreds of demodex. Indeed, the duo were not just ambassadors for humans, but for all kingdoms of life. Meanwhile, scientists are only just beginning to catalogue and understand the full diversity of the human microbiome, and there will surely be surprises ahead.

*FOOTNOTE: In a rare case of a technical name that sounds much more interesting than the common name, parasitic worms are also known as helminths. Most hang out in (and of) the large intestine, but others favour blood vessels. The tiny flatworms that cause schistosomiasis are particularly nasty. They infect some 25 million people each year and kill up to 200,000.

# Most humans contain very little metal

I once had the opportunity to tour a crematorium. I would recommend the experience. The disposal of human bodies was more intriguing than I had imagined. Did you know, for example, that cremation ovens are heavily shielded to guard against explosions? In some cultures, it is traditional to place a bottle of vodka into the coffin. As the temperature rises, the bottle becomes highly pressurized and eventually explodes – often to the surprise of mourners. At least, that's what the technician told me.

I was also shown a store of leftover prosthetics. Even the high temperatures of a cremation oven are insufficient to melt titanium hip replacements or steel pins. Relatives may claim these leftover remnants of the deceased but often they stay in storage for a year at the crematorium. Artificial implants are now common, particularly in the elderly. Meanwhile, most of us will have a dental filling or two, often containing metal. But does metal ever occur naturally in the body? The answer is yes, and in some abundance.

The most familiar is perhaps iron\*. You are home to about 4g ($^1/_8$oz) of the stuff – a couple of peanuts. Still more common is magnesium. This

---

\*FOOTNOTE: Just to be clear, elements like iron and magnesium do not sashay around our bodies as shiny clumps of metal. Rather, they tend to circulate as individual ions – atoms of the element that have lost electrons, such as $Fe^{3+}$ and $Mg^{2+}$. These ions are often surrounded by water or bound to small molecules and proteins.

metal serves as a catalyst, speeding up biochemical processes, including the synthesis of DNA. You have 20–25g (approximately 1oz) of magnesium spread throughout your body. If you could collect it all together (don't – you'd die), it would weigh about the same as two Oreo cookies.

By far the most common metal is calcium. It is a vital component of bone, and also plays a leading role in nerve signalling, among many other jobs. Usually found as the $Ca^{2+}$ ion, it is the fifth commonest element in the body behind oxygen, carbon, hydrogen and nitrogen, and makes up about 1.5 per cent of body weight. Over 1kg (2¼lb) of calcium can be found in the adult body. In more affable terms, your calcium stores weigh about the same as a bottle of wine.

What else can we discover lurking in our chambers and crevices? Potassium is considered a metal by chemists, and it's also present in tangible quantities. Like calcium, it plays dozens of roles in the body, from nerve signalling to hormone secretion. Potassium makes up about 0.2 per cent of body mass, which we can equate to 120–150g (4–5¼oz) in a typical adult. To continue the ludicrous food comparisons, that'd make

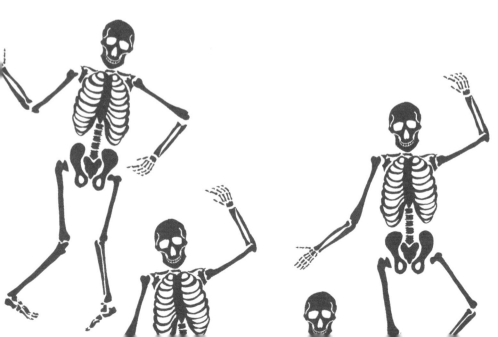

for a decent-sized, if explosive, banana. Sodium is the only other abundant metal, summing to about 100g (3½oz) – a typical pot of noodles before the water is added.

Several other metals are present in our bodies, but only at quantities that could barely form a garnish. Your intrinsic zinc adds up to just 0.2g, though it is essential for life. Copper, chromium, molybdenum and manganese are also vital but scarce. Lead and mercury play no known biological roles, but stowaway particles hide within us all. Even small quantities of silver and gold are there for the taking. The amounts are so scant that you'd have to plunder the bodies of every human in England to get enough gold for a wedding ring – hardly a romantic gesture.

If you broke a body down into its constituent elements, and then sold them at market value, they'd be worth about $150, with potassium being the biggest money-spinner. Any ghoulish marketeers would be better off selling intact organs and fluids on the black market. One old, and slightly dubious calculation from *Wired* magazine put the value at $46 million – assuming you could isolate the individual proteins and hormones as well as organs.

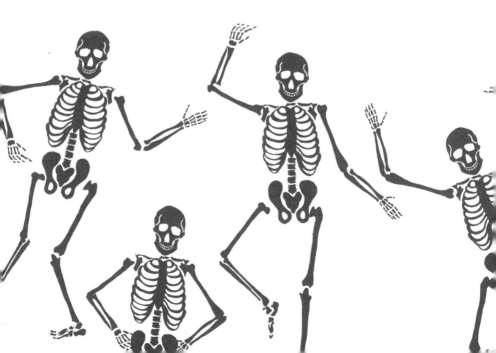

# Skin cells are the main component of household dust

What percentage of the dust in your home is made up of dead skin cells? Let's ask the Internet ...

95 per cent (Yahoo Answers)
80 per cent (*The Vacuum Cleaner: A History*, by Carroll Gantz)
75–90 per cent (Liveabout.com)
70 per cent (Science Shorts)
50 per cent (Graphs.net)

Conclusion: 62 per cent of all statistics are made up on the spot.

It's true that humans shed skin cells at quite a rate: perhaps 10 million a day flake off. But skin cells are tiny, especially when dead and desiccated. Even if you live in a crowded flat, and all of you stayed inside all day doing nothing but rub each other's skin, your cells would only make a minor contribution to the household dust.

If we ask the scientific journals, rather than the Internet at large, we get a more reliable answer. A study in *Environmental Science and Technology* journal (D.W. Layton, P.I. Beamer, 2009, doi:10.1021/es9003735) found that more than 60 per cent of indoor dust comes from outside. This comprises pollen, spores, soil, particulates from construction or traffic, plus insects and their doings. Most of the remaining dust is made up of fibres from clothing, upholstery and carpets. The composition will no doubt vary depending on location, season, presence of smokers or pets, and other factors, but it seems that skin cells are not a major player.

Similar legends turn our own bedrooms against us. It's commonly written that mattresses double in weight every 8–10 years thanks to the accumulation of dust mites and their poop. Pillows are said to be one-third dead skin cells by weight. None of these claims is backed up by scientific analysis (at least as far as I can see). It's time someone did the research and put these undoubted myths to bed.

Dust and dirt are common elements of urban legends. One oft-quoted story – you've probably seen it online – relates to the London Underground. A team of forensics experts supposedly swabbed a row of seats on a Tube train and found the following grim souvenirs: four types of hair (human, mouse, rat, dog); seven types of insect (mostly fleas, mostly alive); the vomit of at least nine people; the urine of four people; human excrement; rodent excrement; human semen. Digging down behind the seats, the team discovered further horrors. Just one row yielded the remains of six mice, two large rats and a species of fungus previously unknown to science.

The damning research is never properly cited and reassuringly impossible to track down in scientific journals. The scientists behind the study work for a non-existent academic department. The report is undoubtedly spurious, spread far and wide by unthinking acceptance that public transport must always be yucky.

# Sex myths

No aspect of the human condition is more open to misunderstanding than sex and sexuality. We could fill another book. Instead, I'll distil it down into a rhyme.

Is it true what they say about men with big feet?
And the longer the pecker, the greater the treat?
Will you ruin your edge if you screw before sports?
No one has tested, or filed the reports.

If you play with your privates to help you unwind,
Will your palms become hairy? Or might you go blind?
Do men think of sex all the time? Well, I'm betting
It's yet to be shown in a clinical setting.

When making a baby you can't sway the gender,
No matter what shapes or positions you tender.
For the chamber of sleep is a realm of false facts,
Like a creature of myth is the beast with two backs.

A condom, a pill or a cap intervention
Are reliable ways to achieve contraception
But there's always a risk, it behooves me to mention
The sure but dull way is for total abstention.

Take care what you read in the blogs and the papers,
Or fake news may temper your sexual capers.
The pumps and the thrusts we perform in the nuddy
Are seldom the subject of peer-reviewed study.

Toilet-seat tales? They're all misconception.
You won't gain a rash or a rogue-jizz conception.
But it's true that sex sells; everywhere, every time.
So ...
To engorge circulation, I've tossed out this rhyme.

# The hungry body

You are what you eat, but only
metaphorically.

# Sugar makes children hyperactive

Contrary to just about everyone's received wisdom, there is no link between the amount of sugar a child eats and his or her level of excitement. There are many good reasons for not overloading a diet with sugar. Avoiding hyperactivity is not among them.

The link, or lack of, has been tested in at least a dozen careful trials. These have looked at sugar in sweets, chocolate, drinks and natural sources. In no study did children on the sugar-free diet behave any differently to their unsweetened playfellows. One study (D.W. Hoover et al., 1994, doi:10.1007/BF02168088) even looked at the effects on parents. Those who thought their children had been given sugary drinks (even when they hadn't) tended to rate their offspring as more hyperactive than normal. In other words, it's a self-fulfilling prophecy.

While we're on the subject of sweet stuff, you might have heard another story concerning the potency of sugary drinks. According to legend, a tooth left overnight in a glass of cola will dissolve to nothing by morning. Imagine what it's doing to your innards! The disappearance is sometimes attributed to the sugar, sometimes the acid, sometimes a combination of both. In fact, the tooth will be largely unaffected by its fizzy dunking. You'd have to leave it to soak for many weeks before any breakdown would be apparent. Further, the tooth would dissolve even quicker in acidic fruit drinks like orange juice. It's one of those made-up 'facts' with little basis in reality, but which we are rarely in a position to test (unless you happen to have a child of a certain age, or regularly get into bar brawls). The myth has such teeth that Coca-Cola even devotes a section of its website to dispelling it.

# Alcohol kills brain cells

Drinking excessive alcohol is never a good idea. Alcoholics face a higher risk of liver and heart disease, among other ailments. Under the influence, you might fall, get into a fight, or generally act like a pillock. And then there's the hangover. Despite the risks, many of us are guilty of occasional over-indulgence. Are we giving our brains a hammering with all that alcohol?

Common wisdom would have us believe that alcohol is the destroyer of brain cells. Some even claim that hangovers are caused by the alcoholic poisoning of neurons. There is a fermented grain of truth to the notion, but – like a hopeless drunk – it doesn't fully stand up.

Any alcohol we drink is broken down in the liver. The organ is effective, but slow at its task. Most people who drink alcohol do so quicker than the liver can handle. That is kind of the point. The surplus alcohol circulates in the blood and passes through the brain. It doesn't destroy the brain cells, but it does interrupt the signalling between them. This leads to slurred speech, lack of inhibitions, dizziness, and all the other wonderful, lamentable and laughable effects we associate with drunkenness. Just as a bad storm might interfere with your TV signal, but won't damage your television, so a night of moderate drinking will only temporarily affect the brain. That said, heavy drinking, especially over many years, can lead to brain damage by other mechanisms, including resulting Thiamine deficiency.

The hangover the following day, then, has nothing to do with the death of brain cells. It's all about dehydration. The shrewd boozer will take a glass of water between each alcoholic drink to avoid a crushing aftermath.

# Eating before swimming causes cramps

Never swim on a full stomach. Doing so might cause cramps, and you will most probably drown in agony. It's a well-known and terrifying warning. As a keen swimmer through my formative years, I never once questioned the advice. My school swimming classes were always mid-morning, to avoid a postprandial dip. Later, as a (somewhat unusual) undergraduate, I'd rise at 6am three times a week for my early bird swim, but I would never eat breakfast beforehand. It's slightly galling, then, to learn this 'fact' is not true. You are no more likely to suffer cramp while swimming if you've just eaten a heavy meal.

Like many other myths we've encountered, this one is underpinned with reasonable logic and half-truths. If you've dined well, then your gut is going to be busy. Digestion needs energy. This is supplied by the blood, which must be diverted away from your limbs and extremities and recruited for digestion. If you attempt any vigorous form of exercise, your limbs can't cope and you get cramps.

That last bit doesn't ring true, though. If you break out into a front crawl, the blood flow will increase to your arms and legs whether you're midway through digesting a meal or not. The body treats this as a fight-or-flight situation, prioritizing blood flow to the limbs over less urgent tasks like processing your last meal. It makes sense from an evolutionary standpoint. A caveman startled by a lion would have a chance of fleeing if his legs were primed; less so if his body were prioritizing a recently swallowed auroch steak.

Cramps are the sudden, involuntary tightening of a muscle, often in the leg. They are painful, but usually short-lived and harmless. If there's one place you really don't want to get cramp, though, it's in deep water. Losing the use of a limb while under great pain is not a winning combination for a weaker swimmer. What causes cramps is still something of a mystery. They are most common in the elderly, those in late-stage pregnancy, or in people on cholesterol-lowering medication – but anyone can be stricken. Dehydration, inebriation or skipping a warm-up might increase the risk. But there's little to suggest that heavy eating is a factor.

If anything, it's a good idea to eat a small, carbohydrate-rich snack not long before a swim, to ensure you have plenty of energy to draw on. A balance must be struck. Eating a double cheeseburger and fries with a side portion of blancmange will give you plenty of energy reserves, but it is not going to make your swim any easier. The reduced blood flow to the stomach can cause feelings of nausea. The horizontal position of swimming also increases the risk of gastric reflux and heartburn. Plus, of course, you're going to weigh slightly more, and therefore be more sluggish, than before you ate. You might feel stomach pains, you may swim slower than usual, you might even be sick in the pool, but you will not be any more likely to get cramp.

# Pregnant women are eating for two

'That's one sausage for me, and one for baby.' Tempting as it may be to double up the portions during pregnancy, the idea that mothers-to-be are 'eating for two' is misleading. The foetus does, of course, need sustenance. It takes in supplies through the umbilical cord. But its diminutive size and idle lifestyle have low energy requirements.

The needs are so low that a woman in her first trimester should have no need to increase her food intake at all (although supplements of folic acid and vitamin D are recommended by many health authorities). Even by the third trimester, a woman of average build and good health needs only around 200 extra calories per day. That's about a 10 per cent increase in food intake. Those expecting babies are eating for $1^1/_{10}$.

Pregnancy carries a raft of other myths. A high bump does not mean 'it's a boy', while terrible morning sickness does not herald a girl. The side that the mother-to-be sleeps

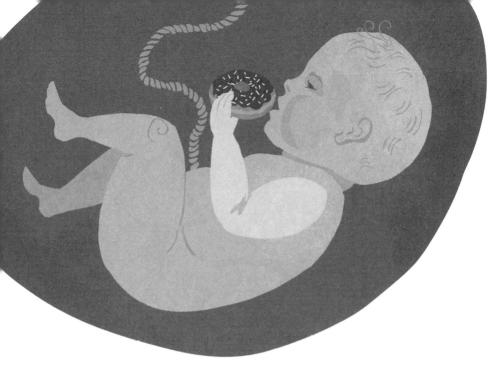

on, the degree of mood swings, the condition of hair, headaches, acne, cold feet – all have been touted as ways of guessing a baby's sex. None are backed up by evidence. There is no way to tell from the outside without a scan. Finally, you can't induce labour by eating curry. The hot food supposedly stimulates the bowel, which in turn tickles the uterus into action. Again, this is mere hearsay.

# Your tongue detects four types of taste

Cast your mind back to school days. You may recall an early science class in which you coloured in a diagram of the tongue. In that classroom, you probably learned that there are four taste groups – salt, sweet, sour and bitter – and different parts of the tongue specialize in their detection. As I recall, it was a lesson undermined during lunch, when the school dining hall had a competency only in salt.

That model of tasting is woefully over-simplified. For starters (or perhaps main course), a fifth basic taste has long been recognized. Umami is the savoury character often associated with Asian cuisine. It's the tricky-to-define essence of soup or broth, but is also found in mushrooms, spinach and fish. Several candidates for a sixth basic taste have also been proposed. How about piquancy, for instance, the hot sensation of eating a meal containing chillis? Some would say that this constitutes an oral experience more akin to pain than taste, but a case can be made for categorizing it alongside sweet, salt, sour and bitter. Some researchers even think that fat should count as a separate taste.

Another candidate is kokumi which, as you'll guess from the name, is another common property of Asian cuisine. This taste translates as 'mouthfulness' or 'heartiness' – a sensation of greater richness. It is used to explain why slow-cooked meat tastes better than a hastily prepared alternative, for example. Kokumi was first proposed by researchers working at a Japanese food company, so you have to wonder if this recent addition to our tongue's repertoire is firmly grounded in science, or whether other interests are at work. Indeed, the whole concept of taste types, including the traditional four, is a little bit iffy.

One way to define independent tastes is to look for receptors – proteins on the tongue that sense the chemical components of food. Sweet tastes, say, are triggered when a sugar molecule binds to a specific receptor in the mouth, like a key fitting into a lock. Sweeteners like aspartame can also dock with the receptor. Sour tastes are common in citrus fruits. Here, the receptor picks up aqueous hydrogen ions from the acidic juice. Salt is self-explanatory; sodium chloride is sensed. Umami is detected by receptors shaped to accommodate small, near-identical molecules called glutamic acid and aspartic acid. Bitterness is the oddity, though. At least 35 different receptors trigger our experience of bitterness. They surely evolved to help us sense a wide range of plant poisons. This ruins the simplistic theory that the four (or five, or six) taste sensations map neatly onto just one type of receptor each. It's a bit more complicated than that.

Indeed, the perception of taste is very complex, and still not completely understood. The key agents are our taste buds – bundles of receptor cells on the tongue that, when grouped together into papillae, give it that rough feel. We start off with about 8,000 taste buds, though the number diminishes as we get older*. The taste receptors poke out from so-called gustatory cells, which form the business end of a taste bud. Most are on our tongues, but not all. The top of the mouth, cheeks and upper throat are also bud-bearing.

It's a myth to say that we have taste buds for salt, and others for bitterness, sour or sweet. Most taste buds can detect several types of taste. Nor do particular tastes tend to cluster on a particular part of the tongue. You may have learned at school that the front of your tongue is exclusively for sweet tastes, while the back part registers bitterness, with sour and salt down the sides. This is mostly bunkum. Careful research in the 1970s showed that all parts of the tongue are equally capable of detecting all the tastes – with the exception that the back of the tongue is slightly keener at sensing bitterness. This misconception will leave a sour taste with those of us who remember colouring in tongue diagrams as children.

The tongue might be the captain, but perception of taste is a team effort. Nose and eyes play an important part in enjoyment of food. This is why you will find it difficult to distinguish orange juice and grapefruit juice if you sample them while blindfolded. And we're all familiar with the loss of taste whenever the nose gets blocked with a cold. The texture and temperature of food – perceived through touch sensations – are also significant.

---

*FOOTNOTE: The number varies greatly between individuals. On average, girls and boys have a similar number of taste buds, although girls typically have a more sensitive sense of taste. Those with a very sensitive palate are often called 'supertasters' – which sounds like a baloney phrase but does have a scientific basis. Supertasters may have a surfeit of taste buds, but their heightened sensitivity can be a poisoned chalice: even mild vegetables like broccoli and cabbage can taste bitter and unpalatable to the supertaster.

# Five food myths that don't pass muster

**1** Raw food isn't better for you than cooked food. Cooking breaks down tough fibres, aiding digestion. It also kills off bacteria. Some raw food might impart more vitamins, but the benefits are negligible.

**2** Saturated fats are not the work of the devil. For years, butter, cheese and other high-fat dairy produce have been demonized and linked to heart disease. Yet recent studies have found no direct relationship. Sugar is now thought to be the villain of the piece. Speaking of which ...

**3** Honey isn't all sweetness and light. It may contain more nutrients than refined sugar, but it has a similar effect on blood-sugar levels, and comes with the same health risks if eaten to excess.

**4** Organic food is not pesticide-free. These crops still use pesticides, though derived from natural products. These may be better for the wider environment, but not necessarily. Further, numerous studies have shown that organic foods are no more nutritious than the cheaper stuff.

**5** It's a total fib that eating celery burns calories. The logic goes that celery is mostly water and indigestible cellulose, so the act of chewing it uses up more energy than the body can absorb – hence, negative calories. It's true that a stick of celery is very low in calories, but the act of chewing and digesting uses far fewer.

# Body oddities

Farts, fingernails and future humans.

# Future humans will have tiny fingers through texting a lot

'What sort of creature is man likely to develop into as the ages slip along? Whether he will evolve a pair of wings, an extra pair of hands, or a few additional senses, are matters on which scientific speculators do not feel qualified yet to speak; but on one – nay, on two points some of them seem to be agreed. The coming man, it is considered more than probable, will neither have hair on his head nor teeth in his gums.'
*Irish News and Belfast Morning News*, 21 April 1897

As the old quip goes: change is inevitable – except from vending machines. The ages turn and the continents shift, and even the forms of plants and animals are gradually revised. We don't see dinosaurs roaming across our fields, though their descendants the birds still pick away at the corn. Likewise, the distant future is unlikely to harbour cows, dogs, sheep, frogs, gnats, cats, numbats or rats, nor any other species alive today. Each is susceptible to the vagaries of evolution, if extinction doesn't remove them first.

But what will humans evolve into? Indeed, are we still evolving at all? These questions have taxed the mind ever since Charles Darwin laid down the principles of evolution by natural selection in 1859*. On the *Origin of Species* describes how organisms can change and diversify in form over long periods of time. Darwin shied away from discussing the implications for humans, saving that for his later book *The Descent of Man* (1871).

It's clear that humans have evolved in relatively recent times. We notionally became 'anatomically modern humans' around 200,000 years ago, but our bodies are not identical to those far-removed ancestors. If we look at humans from different parts of the world, we readily notice differences in outward appearances, such as skin tone, eye and hair colour, and physical stature. People from equatorial regions typically have darker skin than those from the far north or south, for example. Northern Europeans tend to be taller than those from the south of the continent. Such differences evolved relatively recently in our species.

But the differences are more than skin deep. One well-known example is lactose intolerance. Most Europeans are able to digest milk throughout their lives. People in Asia, by contrast, are typically unable to break down lactose beyond infancy. All humans have the genes to deal with milk. In those who are lactose intolerant, the genes are 'down-regulated' after infancy and they do not produce enough enzyme to split lactose into its constituent parts. Drinking milk makes them feel ill.

The ability to digest milk into adulthood maps neatly onto areas with a long history of dairy farming, as you might expect. Thousands of years ago, very few humans would have been able to tolerate milk beyond early childhood. But some members of the community would have been better at digesting the milk than others. By random chance, they carried gene mutations that kept their lactose-pummelling enzymes circulating for longer. These individuals could drink milk later into childhood. Their supplemented diet came with health advantages, and healthier people have a higher probability of passing on their genes – including the milk-favouring mutation – to another generation. Over time, the favourable mutation would have become common through the

---

*FOOTNOTE: Actually, the idea was presented to a scientific audience the year before, along with similar ideas from Alfred Russel Wallace. But it was the publication of the *Origin* that brought evolution by natural selection to a wider audience.

population. Almost everyone could now consume milk and, what is more, cheese.

Those societies who never kept dairy animals were not exposed to the same process. Individuals might be born with the blessed mutation but, as nobody in the community had access to animal milk, they had no nutritional advantage over their mutation-lacking neighbours. Hence, lactose tolerance never arose.

Another famous example is the sickle-cell blood disorder. The condition is caused by a single mutation to a single gene. It leads to malformed, sickle-shaped haemoglobin in the blood. Those who carry two copies of the mutated gene will suffer from anemia and other ill effects. They usually die young without reproducing. But carriers of only one copy gain some resistance against malaria. These individuals tend to live normal life spans, but are more likely to survive malaria outbreaks, and pass on their protective genes. The mutation occurs relatively frequently in sub-Saharan Africa and other regions where malaria is common. Hence, the spread of the trait is seen as an evolutionary response to the disease. More than 20 other genetic mutations have antimalarial effects and may have proliferated by natural selection.

It's clear that humans have evolved in the relatively recent past, but the question remains: are we still evolving? Scientists and commentators are divided on this. It's very hard to devise an experiment that provides a conclusive answer. Evolution typically takes place over many generations, whereas research grants typically last a few years. That's fine if you're studying bacteria, which reproduce every 20 minutes. With the human life span, it's much trickier to catch evolution in action.

Some commentators have argued that humans are now immune to selective pressure. We have stopped evolving. Our technology, social-support systems and understanding of the world are now so advanced that we can mitigate any challenge the environment throws at us – either as populations or individuals. There is some truth in these lines of argument, but also some flaws.

To start with, not everyone in the world enjoys a comfortable life underwritten by effective healthcare, plentiful food, sanitation and shelter. Millions live in dire poverty. Under these conditions, small genetic mutations could make the difference between life and death. Those more able to survive on a limited diet, or ward off a particular disease, really might be at an advantage and live longer to pass on their genes. We've seen it in the past with malaria and the sickle-cell trait.

Scientists are also finding evidence of evolution even in affluent populations. Modern genomic techniques allow researchers to analyze the DNA of thousands of people. With enough data, they can spot how tiny changes to DNA have spread throughout the population in recent centuries. The effects are subtle but seem to indicate that evolution is still playing a part in shaping our biochemistry.

If we are still evolving, how far might it go? Will the generations yet to come really have tiny, dexterous fingers to help with all that texting? As we become more dependent on computers and robots, will our brains begin to shrink from underuse? Could we really grow wings or an extra pair of hands? Speculation about future evolution is nutritious fodder for columnists. It always has been, as the press cutting from the top of this section attests. Editors love to commission this stuff, but it's largely bunkum. Let's pick on tiny text fingers, in case it's still not clear why.

The idea that our fingers might get narrower and nimbler to help us interface with technology makes superficial sense. The world around us is changing, so our bodies adapt to stay on top. But a little thought reveals why this notion is silly. Most obviously, the current habits of typing, texting and swiping may well be short lived. They are cultural habits, open to whimsy. Who's to say that the next generation of tech will care about fingers? We might command our machines with tongue clicks, foot taps, eye motion or perhaps even thought waves.

More fundamentally, to expect fingers to evolve by natural selection to match our devices flies in the face of how the process works. We can imagine that via chance mutations to DNA, individuals will be born who have particularly nimble fingers, and are able to type more efficiently than their peers. But just because a genetic mutation is useful doesn't mean it will get passed on to the wider population. There has to be some kind of selective pressure.

To see why this is, let's take a wacky example. Imagine that I've got a mutation that allows me to secrete gin from my elbow, at will, and in agreeable quantities. All very nice, and a positive boon at parties. I might pass this trait on to my no-doubt grateful children and thence grandchildren. But it's hard to see how the gin-elbow could ever become common among the general population. For a mutation to spread, it must be passed on to the next generation more readily than the boring old version (in this case, elbows that refuse to secrete gin). The mutation would not increase my chances of survival, make me any more fecund, nor more desirable as a mate (not once the novelty had worn off, anyhow). There is no reason my descendants would be any more successful at breeding than their peers. Hence, the gene would not become widespread, and it could not be considered an evolutionary development. Likewise, an individual with narrow, nimble fingers might be excellent at texting, but the talent is unlikely to lead to greater reproductive success. The mutation will not spread widely.

Our species may one day evolve into something with wings, or four hands, or no bodily hair. Over the sweep of millennia, anything is possible. It is also utterly unpredictable. We do not know what selective pressures our descendants will encounter. Nuclear fallout? Global famine? No major hazards whatsoever? Large populations may choose to live off-world, in very different environments. We cannot predict what adaptations might arise in the face of such pressures.

The effects of natural selection might be unforeseeable, but there are other ways to evolve. Humans have unique and still unfolding abilities to alter our bodies by surgical and genetic meddling. Our descendants may be able to see in the dark, avoid cancer, photosynthesize, swim underwater for sustained periods, or perform any number of other tricks that might be accomplished with genetic tweaking. The body will surely be augmented with technology. A process that began long ago with spectacles and dentures, continues today with pacemakers, robotic hands, and that little vibration your phone gives every time someone wants to speak to you. The interface between technology and biology will surely become seamless. Might we one day see a cyborg society whose individuals are part human, part machine? It seems more likely than a human born with wings and four arms through natural selection.

# Identical twins are truly identical

Twins are becoming more common in developed countries. In the USA, for example, the rate of twins remained at about 19 sets per 1,000 births for most of the 20th century. The rate began to go up in the 1980s, and now stands at about 34 sets per 1,000. The chief factor is IVF. Assisted fertility treatment is increasingly common and, when it leads to pregnancy, has a higher likelihood of producing twins than natural conception.

Identical twins are rarer than non-identicals. It happens when a fertilized egg divides into two separate embryos (as opposed to the fertilization of two different eggs, which leads to non-identical twins). Because they derive from both the same egg and sperm, identical twins carry the same genetic instructions. That is why they look very similar. Yet the term 'identical' is somewhat misleading.

For starters, identical twins are not identical if you take a really good look (ask permission first). They might start from the same material, but development in the womb is open to whimsy. One under-construction twin might experience a slightly different concentration of nutrients or hang about in a contrary orientation. One twin may have a more efficient link to the placenta than its wombmate, and consequently grow larger. Moles and birthmarks can crop up in different places. Even fingerprints differ. The twins may have similar whorl arrangements, but the precise pattern of ridges will not be shared.

In rare cases, identical twins may develop as reflections of one another. Mirror-image twins, as they are known, look outwardly the same, but may have opposite dominant handedness, hair direction and tooth patterns. One twin may even show a condition known as situs inversus, whereby their organs are on the opposite side of the body to normal (the heart, say, would have a bias to the right instead of the left). Mirror twins occur when the fertilized egg splits later than normal, after 9–12 days.

Once they're out of the womb, identical twins will diverge further. The lottery of life will see to that. Each is exposed to unique environments and challenges. The siblings will carry individual wounds and scars and achieve

particular levels of fitness. They may be different heights, or weights. One may choose to smoke, eat fatty food or take drugs, while the other does not. Such habits can have a marked effect on gene expression – the process by which the body makes new proteins by 'reading' the DNA blueprint. While the twins have the same DNA, it might be operating in different ways thanks to these so-called epigenetic factors. All this is before more conscious decisions about appearance, such as hair style, piercings and tattoos. Two adult twins may look dissimilar.

Even at a genetic level, identical twins are not truly identical. Both begin life with the same DNA, but that template can change over the years. It's like a piece of website source code that gets shared among developers. Each will introduce tweaks and changes that mean the once-identical copies no longer match. DNA can be scrambled by the presence of carcinogens, or from exposure to certain types of radiation. These cause random mutations to the DNA or introduce errors when DNA is copied. Such mutations accumulate as life goes on, meaning that your genetic makeup at death is subtly different to the one with which you were born. Most mutations have no visible effect, but some may lead to diseases or a physical mark on one twin and not the other.

Identical twins, then, are never truly identical. The term is flawed, but universally understood. The scientific and more accurate alternative – monozygotic twins – is too much of a mouthful to catch on.

# Humans can spontaneously combust

*Bleak House* by Charles Dickens is a novel wheezing with fog and smoke. The strangest wisps of all come at the end of Chapter 32, when the gin-sodden rag merchant Mr Krook is found burnt to a cinder in his own home. It is a case of 'spontaneous combustion'. The unfortunate man had self-ignited, with no external flame or heat source.

In his introductory notes to a later edition, Dickens reveals his inspiration. He cites the story of Countess Cornelia de Baudi of Cesena, Italy. In 1731, the Countess was found dead in her bedroom. Or, rather, she was found dead about her bedroom. The lady didn't succumb to simple burns, she had been all but consumed. Only her lower legs, three fingers and part of the skull remained. Everything else had been reduced to ash and an odorous oily soot, which coated everything in the room.

How can a human body – which is mostly water – burn so completely without the presence of an oven or furnace? Under the right circumstances, a body can behave like an inverted candle. Melted fat can soak into clothing, which then acts like a wick. Indeed, the phenomenon is known as the 'wick effect'. Victims are often left with their hands and feet intact.

These body parts contain relatively little fat and are least likely to be covered with clothing – hence they are less susceptible to the wick effect.

Cases of extreme combustion date back centuries, and still occur today. These terrible infernos are sometimes put down to spontaneous combustion – as though the body caught fire of its own whim. The idea is now widely discredited. No plausible mechanism has ever been described. Further, if the human body really could catch fire by itself, then why don't we see it more often? How come it's never been witnessed or filmed in progress? It is far more likely that these fires are started by an external flame source – like a lamp or cigarette – that is itself consumed by the fire. Even so, there is still some belief in the phenomenon. As recently as 2010, an Irish coroner recorded spontaneous combustion as an official cause of death when no other explanation was forthcoming.

Humans and flames rarely go well together. Yet there are singe-free ways to play with fire. The idea of walking across burning coals may send a shudder of fear up the spine. It seems like an impossible act, and has been tied up with rituals and mysticism for millennia. There's nothing supernatural going on. With a little confidence, anybody can do it. So long as you walk at a steady pace, your foot will not be in contact with the coals long enough to cause a burn. Running is actually more hazardous. The force of each step would take the foot deeper into the coals, beneath the relatively cool, ashy upper layer.

Fire eating is a lot tougher. The performer extinguishes the flame by closing the mouth, cutting off the oxygen supply. This is fraught with hazards. Magician Penn Jillette described his traumatic initiation into the art, as a teenager: 'I practiced all afternoon and burned the snot out of my mouth and lips. My mouth looked like wall-to-wall herpes sores, with cartoonish, giant teeth glued to my lips.' With practice, though, it's perfectly possible to douse a flame in the mouth without resorting to illusion.

# Fingernails and hair continue to grow after death

If you ever find yourself among the ranks of the undead, you might want to track down an accessory stall. The typical movie zombie has long matted hair that really should be tied back with a band. It just gets in the way of a good mauling. Zombies also sport long fingernails, verging on claws. Both are symptoms of passing over. For when we die, our hair and fingernails continue to grow.

This well-known piece of hearsay is as mythical as zombies themselves. It isn't physiologically possible for any tissue to grow after death. Hair and nails are made chiefly of keratin, a tough, fibrous material. For them to grow, the constituent cells would need to divide. This can only happen if there is a supply of raw materials and energy, and that requires oxygenated blood flow. When the heart stops, so too does blood flow. There is no way the cells can divide.

Even so, there are many accounts of exhumed corpses that look hairier than expected. There may be two reasons for this. First, the body dries out in the days following death. Skin contracts and begins to break down. Hairs, and particularly stubble, stand out against the diminished flesh. Second, the keratin that makes up hair and nails takes much longer to break down than other tissue. It is the most obvious feature on a long-withered corpse.

# Shaving causes hairs to grow back faster, or rougher

No matter how close you shave, you won't be touching living tissue. The part of each hair that is alive and growing, called the follicle, is subcutaneous – that is, within the layers of skin. When a razor slices the hair close to the skin, it is cleaving a path between cells that are no longer alive.

Dead hairs don't talk. No message is sent back to the hair root to say, 'Hey, look, I've just been decapitated.' As far as the follicle is concerned, you might have chopped off the whole visible hair, just the tip, or nothing at all. Consequently, there is no difference in the quality of hair that eventually grows back. Nor does it shoot out with extra speed.

Experimental evidence, such as there is, seems to back up this chain of logic. Studies as long ago as 1928 found no difference in growth rate following shaving. Another study in 1970 found the same, although the sample size was tiny (five healthy white men), and only applied to leg hair. It seems there is scope for further research, although few would expect to see any effect. After all, if hair really did come back thicker or coarser after shaving, I'd have a chin like a pan scourer by now.

Where does this shaggy-dog story come from, then? It seems to fit with our experiences. Stubble on our chins or legs feels much coarser than locks that have been allowed to grow. This is simply because the hair closer to the root is a little bit thicker than hair that has grown out. It's no coarser than it was before the shave, but the removal of the tapered ends makes it seem so.

Stubble is also a bit darker than hair that has grown long. Light exposure and chemical pollutants can cause bleaching – the longer the hair has been exposed, the greater the degree. Someone who goes from flowing coiffure to crew cut will be left with only the darker sections of hair, which can add to the impression of coarseness.

Waxing hair *does* get to the root. The entire hair is yanked clean out, along with dozens of neighbours. This is painful (I'm told), but leaves the skin clear of hair for several weeks. When it does grow back, the hair is usually of the same thickness as previously. Multiple waxings can lead to finer hair, as root structures become damaged. It certainly won't make the hair grow back thicker or coarser, as is sometimes rumoured.

# Some people just don't fart

The Queen farts. It's true. Few will ever witness the august bottom burp, but Her Majesty must parp at least a dozen times a day. All humans do. There is no avoiding it. Most of the noisome gas is created as a byproduct when bacteria in the gut break down food. The rest comes from swallowed air or perhaps carbon dioxide from fizzy drinks. Speaking of which, a typical daily gas yield would fill a 2-litre cola bottle – an experiment most schoolboys will have contemplated at some point.

No human (or other creature with a gut) is exempt. We all produce bubbles of gas. You can try to hold it in, and you might think you've succeeded, but the vapours will escape sooner or later. The unspeakable gas may sneak out slowly and stealthily – or might issue forth with an uncontainable pop. But farts do not get reabsorbed or go away.

Not all farts are obvious, though, even to the dealer. The main emissions are methane, hydrogen, nitrogen and carbon dioxide, all of which are odourless. That distinctive smell comes from hydrogen sulfide and other sulfurous compounds, which make up just 1 per cent of a typical emission. If you've not recently eaten sulfur-heavy foods, your exhaust may be almost odourless.

Methane and hydrogen are both highly flammable. The first is the main constituent of natural gas – the stuff we burn on our stoves. The latter was used in early airships. Its tendency to ignite caused many tragic accidents such as the *Hindenburg* and *R101* disasters. Not surprisingly, then, it's no myth that farts can be lit. I wouldn't encourage anybody to try it for themselves, but copious evidence can be found on sites like YouTube.

# Dead bodies are inert

'Die, my dear? Why that's the last thing I'll do!'
Groucho Marx

Actually, Mr Marx, it's not. The last act that many of us perform is a fart. In the hours and days after death, gas trapped within the corpse finds its way out. There are two options: the mouth or anus. Any 'fart' will be silent as

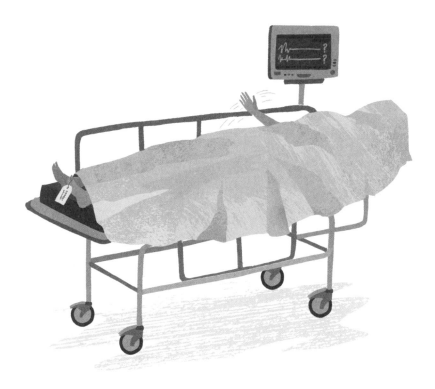

the sphincter muscles will be too relaxed to cause an audible parp. It's not just gas that escapes. If the bowels and bladder were full, the contents will eventually emerge. Many people do not witness their own final defecation.

Dead bodies are more active than we might think. As the corpse is broken down by bacteria, decomposition gases are produced. The side effects can be alarming to anyone not used to watching bodies decay (that is to say, most of us). When the gas escapes from the mouth, the body may appear to be sighing. The release can cause the body to move, as the dead weight shifts on its centre of gravity.

Sometimes, the gases get trapped. A corpse may bloat up as the pressure builds. At the critical juncture, the gases and liquids of the decomposing stomach can make their way up the oesophagus and pour out like vomit. Women who were pregnant at the time of death may even 'give birth' to their dead child. The phenomenon is known as 'coffin birth' and is caused by pressure from decomposition gases pushing the foetus out.

Rigor mortis usually sets in between two and four hours after death. Contrary to popular notions, this does not cause limbs to move, nor does the corpse adopt a lifelike pose. The tightening of the muscles merely makes the limbs stiff and difficult (for someone else) to reposition. Some corpses appear to be grimacing as the jaw muscles tighten. Under rare circumstances a dead man may get an erection. Those who die with injuries to the cerebellum (the lower back part of the brain) or spinal cord, such as hanged men, are the most likely to do this. Cue gallows humour about 'stiffs'.

# An A–Z of quackery

Modern medicine has its shortcomings but it is, at least, based on a scientific understanding of the human body. Doctors can only prescribe medicines that have been rigorously tested and held up to scrutiny by other trained professionals. But there are other systems of healing and wellbeing that do not have a firm scientific basis. Some, like acupuncture, perpetuate ancient traditions; others, like homeopathy, were developed in comparatively recent times.

The techniques of 'alternative' medicine are spectacularly varied – from sticking needles in the body to drinking urine. All have two things in common: anecdotal evidence of effectiveness and zero scientific evidence of effectiveness. Those who pursue alternative therapies or diets are often convinced of the benefits, yet laboratory studies seldom find a link. To those of a rational mindset, alternative medicine is nothing more than quackery. Here, then, is an A–Z of some of the most common examples.

**Acupuncture:** The insertion of needles into the body at specific points (see Qi, page 108) in the belief that it can heal or reduce pain. Scientific tests of the practice have shown no benefits that can't be put down to the relaxing environment and placebo effect. So many points, so little point.

**Biorhythms:** My first computer (an Acorn Electron, if anyone else is old enough to remember them) came with a bundle of free software, including something called Biorhythms. All it needed was my date of birth, and it would spit out a raft of predictions about my current mood and health. The calculation supposedly took into account the natural rhythms of the body: a 23-day physical cycle, a 28-day emotional cycle, and a 33-day intellectual cycle. By working out where I was in each cycle, based on the number of days since birth, the algorithm could make predictions about my wellbeing. At the time, I dismissed it as a fun showcase of computing power. But it turns out that the software was based on a genuine theory that still has many adherents. There is no evidence that such cycles exist (other than a tenuous link to the menstrual cycle in females), and the theory has zero scientific credibility.

**Chiropractic:** A chiropractor manipulates and massages the spine in an attempt to relieve pain or cure illness. That sounds plausible. However, the practice was borne of mumbo-jumbo and mysticism, which still pervade some parts of the profession. One of the central beliefs – vertebral subluxation – supposes that tiny misalignments in the spine are connected with ill health. Nothing in conventional medicine supports this idea. Further, careful and critical reviews of chiropractic have found no benefit for many of the conditions for which it is offered. Some studies have shown minor improvements when chiropractic treatment is applied for lower back pain and related discomfort, but the practice is of little demonstrable benefit for other ailments.

**Dietary supplements:** According to the US National Institutes of Health, Americans spend over $30 billion a year on nutritional supplements. The list of products is bewildering. The same source estimates that some 75,000 supplements are now available in the USA, including vitamins, fish oils, probiotics and other products. Many people have a good reason to take such supplements. Women are strongly encouraged, for example, to take folic acid during pregnancy, which helps to reduce the chance of birth defects. Those confined indoors through illness or for other reasons can benefit from vitamin D supplements. Many illnesses are caused by deficiencies in one vitamin or another and popping a pill can be a big help. A large wad of that $30 billion, though – and much more overseas – is misspent. Supplements are usually unnecessary for those who eat a reasonably balanced diet. We might as well spend our hard-earned cash on magic beans – at least they'd offer a good serving of iron and B vitamins.

**Ear candling:** The idea behind ear candling sounds vaguely convincing. Stick a hollow candle in your ear hole, light the wick, and the negative pressure from the flame will draw out ear wax and other impurities. All this while filling the air with a pleasant aroma from the scented candle. There are three downsides: you look silly, it doesn't work and you can end up with burns. Several studies have shown no effect (for example D.R. Seely et al., 1996, doi:10.1097/00005537-199610000-00010), while also pointing out the dangers.

**Flossing:** It's hard to believe that something as mainstream as flossing might be futile, but that may be the case. A good floss will dislodge particles of food and reach areas beyond the bristles of a toothbrush. The American Dental Association and the UK's National Health Service both recommend we do it. Yet there is precious little evidence that people who floss have healthier teeth and gums. The few studies that have tested flossing are flawed. Some were too small, while others were funded and directed by floss manufacturers. This is not to say that flossing is ineffective, just that there isn't a strong evidence base to support it.

**Glucose drinks:** When undergoing a heavy workout, the body needs to take on more energy to keep up performance. Glucose drinks can provide

that fuel in the form of glucose – a simple sugar, which the body can readily harness. The trouble with glucose drinks is that they contain staggering quantities of sugar – far more than most physical exercise could need. Some regular-sized bottles of 'sports drink' contain 16 teaspoons of the sweet stuff. The odd slug of energy drink now and then isn't going to cause any harm, but people who consume sports drinks too regularly are more likely to put on weight than to lose it. A recent small-but-intriguing study suggested that sucrose – the sugar we all keep in the cupboard – may be more effective than glucose for energy replenishment. A drink of sugary water was shown to outperform a glucose drink in a study of endurance cyclists (see J.T. Gonzalez et al., 2015, doi:10.1152/ajpendo.00376.2015).

**Homeopathy:** This is an alternative therapy based on the notion that like cures like. A common example is insomnia. The problem might be alleviated by taking a diluted sample of something that causes sleeplessness, like coffee. Leaving aside the counter-intuitiveness of this idea, homeopathy has another big oddity. Its treatments are very diluted, sometimes up to 400 times. If you crunch the numbers (and homeopathy is so nearly an anagram of 'phooey math'), you get a startling result. The preparation is so watered down that not a single molecule of the original substance is likely to be present. It is simply pure water.

How, then, can a homeopathic preparation have any effect on the body? It can't, according to any credible scientific theory. Nor have homeopathic treatments ever shown much promise when put under laboratory scrutiny. Yet the practice remains popular. Plenty of people claim to feel better after taking homeopathic medicine. Perhaps they are benefiting from the bedside manner of the homeopath. A typical consultation lasts far longer than a visit to a conventional doctor. The homeopath takes time to get to know the patient and their lifestyle ('ooh, empathy', to deploy another anagram). This positive experience may enhance the placebo effect and help the wellbeing of the patient. It won't cure any underlying diseases, though.

**Ice baths:** Athletes and bodybuilders sometimes immerse themselves in icy water after a vigorous workout. The frigid dunking is thought to reduce swelling and tissue damage by slowing down blood flow; much as a bag of frozen peas will alleviate the pain of a bump to the head, the icy waters can take the edge off a burning muscle. This is all well and good, but there's evidence that the ritual might be counter-productive for those looking to put on muscle. A recent study (L.A. Roberts et al., 2015, doi:10.1113/JP270570) looked at the differences between men who dunk and those who instead 'warm down' on an exercise bike, following strength training. After 12 weeks, muscle strength and mass had increased more in those who favoured the warm-down. Further, biopsies showed that key proteins needed for muscle repair were 'blunted' by the cold-water treatment for up to two days following immersion. The study doesn't argue with the benefits of a shivery dip for pain relief, but it does throw cold water on the idea that an ice bath helps muscle repair and growth.

**Just about anything you read on a moisturizer packet:** I can't claim to be an expert on skincare and beauty products. I do, though, have a degree in chemistry, and another in biochemistry. So, when I read the claims of the skincare industry, I often raise a puzzled eyebrow at some of the terms. Moisturizers, of all the unguents, seem to be the most claptrap-happy. Meaningless phrases abound. 'Rebalance your skin', 'infused with healing oxygen', 'skin-loving ingredients', 'multi-intensive', 'a fragrance that exudes energy' and 'optimizes the efficacy of oxygen levels', are just six examples I found in a five-minute, cursory web search. All of these phrases superficially sound like they've come from a scientist, but the words are vague and of questionable meaning.

Even familiar terms can be woolly. Hypoallergenic, for instance, might lead you to believe that the product is less likely to cause allergic reactions than your typical moisturizer. The thing is, the word has no official meaning and can be used willy-nilly by manufacturers. I bet you can't find a single moisturizer that doesn't claim to be hypoallergenic. Likewise, most moisturizers say they've been 'dermatologically tested', that is, tested for adverse effects on the skin. Well, you'd hope so, wouldn't you? But, again, there's no official standard of meaning here. The manufacturer might well have done a study, but we don't know what exactly they were testing. My car recently went to the garage for its annual check-up. I can now boast that it has been 'mechanically tested', even though it didn't pass.

And then there are the baffling ingredients on these things. Many contain a litany of 'natural' products that reads like a fairies' picnic: mulberry root, tree moss, tonka seed, grapeseed, wild bluebell, honey extract, pomegranate, apricot oil … These are often supported by a stern cohort of sciency-sounding ingredients, as though to add an aliquot of credibility. Look out for aquaporins, retinoids, and, of course, antioxidants.

The (largely unused) moisturizer in my own bathroom contains a terrifying list of over 30 ingredients, including tocopheryl acetate, ethylhexylglycerin and various other mouthfuls that my spellchecker won't allow. Occasionally, you get a fairy-science crossover, like the common ingredient of 'colloidal oatmeal'. One brand advertised on Amazon even contains an 'alcoholic mother' among its ingredients. In any other arena, I'd assume that to be a malicious typo, but with moisturizers, you never quite know.

**King's touch:** For the sake of variety (and, admittedly, because I'm struggling to find a 'K'), here's an historical form of quackery practiced by the highest in the land. The King's touch was the belief that an august fondle could cure disease. The King (and sometimes Queen) was believed to have divine favour, which he might channel into the power to heal. Those most sovereign hands were particularly talented at curing scrofula, a lymphatic disease that came to be known as the King's Evil. The practice was common in England and France. Charles II of England is said to have touched some 92,000 scrofulous people during his reign – equivalent to about one-quarter of the population of London at that time. Obviously, no double-blind clinical trial was ever set up to determine if the monarch's hands really were an effective cure for scrofula. That said, I think we can put the efficacy of King's touch down to scrofula's natural habit of clearing up of its own accord.

**Leaky gut syndrome:** Did you know that your gut can leak? Harmful substances can seep out, causing menace to the wider body. Leaky gut syndrome has been blamed for everything from autism to migraines to multiple sclerosis. Fortunately, it's a fictional syndrome, at least as far as medical science is concerned. Like all pseudoscience, though, it embellishes a lesser truth. The gut wall can indeed become more permeable, allowing undigested food, microbes and other material into the bloodstream. This can be a factor in conditions such as Crohn's disease and diabetes, though a direct mechanism of cause and effect has not been established. Leaky gut syndrome goes much further. Its proponents blame percolating innards for a wide range of diseases. They may diagnose based on scant evidence and offer supplements and probiotics with no proven effect.

**Magnetic healing:** How awe-inspiring it was, as children, to play with invisible forces. Magnets had the potential to do anything. I'm sure I wasn't the only child who ruined a pair of training shoes, stuffing the heels with magnets in a futile bid to invent 'levitation boots'. Magnets have properties that can seem magical, so it's no wonder they've been touted as medical cures. Do you know somebody who wears a magnetic bracelet? The jewellery is supposed to improve blood flow and thereby general health through its magnetic interaction with hemoglobin. This is nonsense. Even an MRI scanner, which uses magnetic fields many thousands of times more powerful, has no observable effect on blood flow. Even if it did, there's no obvious reason why this should lead to better health. Magnetic therapy is usually a harmless con, with no side-effects. Like many quack remedies, however, if it is offered as a cure for serious diseases, such as cancer, it can give false hope and delay professional treatment.

**Neuro-linguistic programming:** I've always thought this motivational technique's name sounds a bit sinister – like some kind of mind probe from a science-fiction show. It's a little hard to describe without slipping into NLP lingo, but in essence, the technique looks at the mental, physical and talking habits of geniuses, and teaches them to us humble non-geniuses. Or something. Look, I'll level with you. I've spent several hours reading up on the subject in preparation for this paragraph and now my head hurts. Someone needs to give me NLP to help me understand NLP. My personal inability to grasp the subject does not mean the technique lacks merit, or that it doesn't work. (That's a common logical fallacy called an 'argument from personal incredulity' – I don't understand the thing, therefore it can't be true.) Fortunately, plenty of research has been done on NLP to help people like me make a decision about whether to give it a go (see, for instance, T. Witkowski, 2011, doi:10.1186/1471-2474-15-286). Conclusion: there is no reliable evidence that NLP works, and its underlying theories have no scientific support.

**Osteopathy:** Like chiropractic, osteopathy has a convincingly scientific name, yet little scientific underpinning. The practice involves the manipulation of muscles and joints to improve well-being, like a massage with medical benefits. There is 'moderate-quality evidence' that it can help relieve lower back pain (H. Francke et al., 2014, doi:10.1186/1471-2474-15-286) and muscle pains after surgery or injury. In that context it would be harsh to describe osteopathy as quackery. However, some osteopaths reckon to cure illnesses unconnected (in conventional medicine) to the muscles. Asthma, migraine, digestive disorders, Parkinson's disease, depression and even colic in babies, have all been tackled by osteopaths. There is no rigorous evidence to support this, nor any plausible biological mechanism by which osteopathy could help with such conditions.

**Psychic surgery:** Remember those cheap conjuring tricks, beloved of uncles at birthday parties, in which a coin disappears, only to reappear behind a child's ear? Imagine doing that with a handful of gore. That's all there is to 'psychic surgery', a widely discredited technique that supposedly removes tumours without surgical instruments. The practitioner presses down on the affected body part, magics up a pool of blood, then pulls out an unappetizing clump of matter. The blood is then politely washed away to reveal no wounds.

Like any illusion, psychic surgery can look convincing, but it is nothing more than nasty sleight of hand. Its roots lie in both Brazil and the Philippines, but credulous audiences have been found elsewhere. Most famously, the entertainer Andy Kaufman sought the help of a psychic surgeon to cure his lung cancer. He died from the disease just months after claiming he'd been cured.

**Qi:** QI (all capitals) is a British comedy panel show in which celebrity guests attempt to out-fact one another. It's jolly good, and one of the key inspirations for these books. Qi, on the other hand, is a 'life force' or biological field said to permeate all living beings – if you wanted a crass metaphor, it might be compared to 'The Force' in the Star Wars films. Qi underpins many aspects of Chinese culture, such as feng shui, traditional Chinese medicine and certain martial arts. I'm sure that belief in qi can help with mental discipline and wellbeing, but it doesn't have much use in a medical setting. As Wikipedia tactfully puts it, 'Despite widespread belief in the reality of qi, it is a non-scientific, unverifiable concept.'

**Reflexology:** Reflexologists believe that the organs of the body are mirrored by corresponding zones on the feet and hands. Gentle pressure to the heel might relieve pain in the buttocks, while a quick rub below the little toe can ease a throbbing in the arm. Practitioners might know their arse from their elbow in this sense, but reflexology is otherwise without any factual basis. It is yet another technique that relies on unverifiable explanations of 'body energy'. A 2009 review drew together all the competent literature on reflexology. It concluded, based on 18 different data sets, that 'The best evidence available to date does not demonstrate convincingly that reflexology is an effective treatment for any medical condition.' (E. Ernst, *Medical Journal of Australia*, 2009, Vol. 191, pp.263–266 – no doi available.)

**Superfoods:** A new superfood comes along every month. Kale, blueberries, papaya and Popeye's original superfood, spinach, have all met the spotlight. Ever more exotic and mysterious foodstuffs like quinoa, goji berries and jujube have all taken their turn as the fashionable food for the health-conscious. The trend is now self-parodic. The latest phase has seen crickets, hemp, snails and even edible clay touted as the new wonder food. One senses that the craze is more about feeding magazine columnists and an Instagram audience than one's own stomach. Unless you happen to like the taste (and I am rather partial to quinoa), save your money and buy local fruit and veg. Most 'superfoods' are rich in vitamins and other nutrients, but they have zero advantage over a diet that contains more mundane greens.

**Therapeutic touch:** Emily Rosa of Colorado holds a Guinness World Record as the youngest person to have a research paper published in a peer-reviewed medical journal. She was just nine years old when her study of therapeutic touch was published in the highly prestigious *Journal of the American Medical Association* (L. Rosa et al., 1998, doi:10.1001/jama.279.13.1005).

Therapeutic touch is a recent pseudoscience, developed in the 1970s, but with roots in ancient tradition. Practitioners claim they can feel and manipulate the human energy field and thereby speed up healing and reduce pain. That would be lovely, save for the fact that humans don't have an energy field that any proper scientist has ever detected. The claims were put to the test by Emily Rosa as part of a school science fair. She asked practitioners to put both their hands through a screen so that they could not be seen. She then placed her own hand a few centimetres above one of them. The practitioner had merely to sense her energy field and declare 'left' or 'right'. The results were damning. The subjects guessed correctly only 44 per cent of the time, a number consistent with pure guesswork. The study, which was backed up with rigorous statistics and repeat experiments, undermined any scientific basis of therapeutic touch.

**Urine, drinking of:** Have you ever considered drinking your own urine? It's unlikely to hurt you. The golden stream is essentially sterile when fresh. It's also of little value – even if you're trapped for days in a lift with nothing else to drink. Urine is high in salts, and unlikely to quench your thirst. Any unexpected effects, good or bad, from drinking urine are hard to pin down. Recruiting enough volunteers for a water-tight clinical study would be a wee bit tricky.

There are, however, groups of people who advocate the drinking of urine for therapeutic reasons. It is particularly common in India, where former Prime Minister Morarji Desai was a noted champion of a daily quaff. Advocates tend to recommend the practice for harmless cosmetic improvements, like hair loss or skin tone. At the more pernicious end of the spectrum, it's easy to find websites that claim urine therapy can cure mental illness, HIV/Aids or cancer. Needless to say (yet again), there is no reputable scientific evidence that drinking urine can have any positive effects on health.

**Vacuous fad diets:** Most people know that there is no miracle solution to losing weight. Even so, tips for shedding the pounds are among the most common forms of health advice in the media. Most of the time, these are harmless variations of 'eat fewer calories and do more exercise'. This will work if you have the willpower. And then there are the fad diets. Fad diets typically share three properties: (1) a catchy title, usually named after a fruit or a fruitcake; (2) celebrity endorsement; (3) no basis in science. There are grapefruit diets and lemonade diets; meaningless 'detox' diets that do not and cannot remove toxins; paleo diets that would have us eat like our hunter-gatherer ancestors while not getting anything like as much exercise; illogical 'alkaline diets', which supposedly balance the pH levels of the body, but do no such thing.

Some fad diets may even be harmful. Followers of the bizarre 'cotton ball diet' soak said item in fruit juice before swallowing. The cotton ball gives feelings of satiation without contributing any calories. It can also leach toxins and potentially block the bowel. The so-called 'Sleeping Beauty' diet insists that you sleep for longer to reduce meal opportunities. Some followers take sedatives to make it last. A whole book could be written debunking this stuff but I just don't have the appetite.

**Water:** Water would seem a daft inclusion on a list of quackery. It's one of the essentials of life. Water makes up more than half our body weight. Yet the benefits of drinking water are often overplayed in western societies. Health authorities typically recommend we take a couple of litres (3½ pints) every day. This doesn't mean we need to knock back two litres of pure, workaday water. The body is largely agnostic to the source. Food provides some of the liquid we need, especially fruits, vegetables and anything soupy. Soft drinks, fruit drinks and hot beverages may cause health concerns in other areas, but as sources of water, they're fine. The oft-quoted idea that we 'need' to drink eight glasses of water a day is simply not true (and confusing, given we all have different sized glasses). So long as you have pale pee, and don't feel thirsty all the time, you're probably taking in enough water. Do as your body demands.

**'X' causes cancer:** Everybody knows someone who's had or has cancer. About half of us will be diagnosed at some point in our lives. It's a theme of universal interest. Consequently, 'cancer' is a good word to put in a headline if you're a news editor and want to attract readers. Try doing a search right now for the phrase 'cure cancer' or 'causes cancer'. At the time of writing, I get such insights as 'British sausages DON'T cause cancer', 'Jet lag could increase your risk of cancer', 'Metro travel can cause cancer', and 'Assam minister says sins cause cancer'.

Headlines like these are almost always spurious. Often, the story is reporting on a scientific study that has found a link between a particular agent and a particular type of cancer – but only in cell culture or mice or some other non-clinical setting. Journalists or their editors often twist limited research findings into monumental certainties. 'In our study, X was associated with a slight increase in skin cancer in mice' is often translated as 'X causes skin cancer!'. When the research is relevant to humans, the results are rarely clear-cut. It is fiendishly difficult to link a given food or substance to a particular disease. How, for example, can you demonstrate that people who eat lettuce every day run a lower risk of developing cancer? Is it down to the salad itself (a causal relationship), or that such people probably lead healthier lives in general (a correlation)? How you might collect such data is a fascinating subject in its own right, and I refer you to Ben Goldacre's excellent book *Bad Science* (2008) for erudite discussion on the subject. I'm betting that the Assam minister isn't basing his assertion that sins cause cancer on a rigorous clinical trial or epidemiological study.

**Yoga:** I know many good and sensible people who practice yoga. It is an excellent, all-round workout that improves suppleness and strength without the high-impact shocks common to many sports. It is also suffused with dubious health claims and mysticism – what some have called 'yoga woo'. Talk of 'energy fields', 'aligned chakras' and 'cleansed internal organs' may help set a mood and environment, but medically speaking it's all unverifiable gobbledygook. That's not a problem until people in power start pushing it as fact. That happened in 2016, when a member of India's parliament (not the same guy who advocated urine drinking) claimed that certain forms of yoga could prevent and cure cancer. This is a dangerous thing for a government official to say. The risk is that people might forego effective treatments in favour of unproven variations on bending and stretching. His promise of scientific evidence has yet to be published at the time of writing.

**Zapping the body:** 'ELECTROPATHIC BELT GUARANTEED to RESTORE Impaired VITAL ENERGY, Invigorate the Debilitated Constitution, Stimulate the Organic Action, Promote the Circulation, ASSIST DIGESTION, and promptly Renew that Vital Energy, the loss of which is the first symptom of decay' – so reads an advert from the Western Times, 9 March 1889.

It's not hard to find adverts for electrotherapy in Victorian newspapers. The public were agog at the power of electricity. Very few people had electricity supply to their homes, and the energy source retained a magical quality. Quacks lined up to exploit the credulity. The man selling electropathic belts above was the almost too-aptly named Dr C.B. Harness, a 'medical electrician' who used a barrage of pseudoscientific words and random capitalization to promote his electrically stimulating belts. These could cure anything. The adverts list epilepsy, gout, paralysis, indigestion and constipation, among many other ailments that might succumb to electric engirdlement. Forget your drugs and crutches, one zap from the magic belt and you'll be right as rain.

Shocks and currents have been put to medical use ever since electricity was first tamed in the 18th century. The practice has yielded useful techniques such as electroconvulsive therapy for certain types of depression, and deep-brain stimulation for neurological disorders. It's also been fertile territory for cranks like Dr Harness. Somewhere between the two are the raft of interventions that have some kind of credibility despite a dearth of strong evidence. Chief among these is transcutaneous electrical nerve stimulation, which is used to relieve pain by passing an electrical current through the skin. Many people swear by TENS, but good-quality evidence for its effectiveness is lacking. Another controversial technology is the 'thinking cap'. These devices sit on the head and stimulate neurons in the brain with an electric current. Can they really boost your mental prowess? The evidence is shaky.

# The body compromised

Misconceptions about illness, injury and ailments.

# You'll catch a cold if you go out in the rain

Call it Jane Austen syndrome or the Dickensian flu: the tragic lover caught in a rain shower is a familiar cliché of the classic novel. Soaked to the skin, they will either be bedridden for the next 12 chapters, or die from the chill. The idea is not limited to the 19th century. Even preschool show *Peppa Pig* sees a rain-sodden George develop a sneeze.

It might be a melodramatic way of moving the plot forward, but the idea of a lengthy illness caused by a drenching is exaggerated. Colds and flus are viral infections. You need to be exposed to someone or something with the virus before you will catch it yourself. Raindrops do not carry the virus. Nor do chilly winds. You cannot catch the flu by being wet or being cold, or going outside with wet hair, or running through a rainstorm following an awkward tryst.

There is an 'however' here, which allows the myth to claw back credibility. Frigid temperatures cannot in themselves give you a cold, but they could hasten its onset if you already harbour the virus. Exposure to cold causes the blood vessels in our extremities to contract. This restricts blood flow to areas affected, which may include the nose and throat. This, in turn, means fewer immune cells can reach those areas. Cold exposure, then, could reduce the efficiency of an immune response to a virus that's already in our airways. The evidence remains inconclusive, but it's a persuasive idea. Another factor is that cold or wet weather tends to keep us indoors for longer periods. The warm, confined conditions, close to other people and the germs they carry, are perfect for transmitting the virus.

# Vitamin C prevents colds

Perhaps you don't believe everything you read in the last section. You've been out in the rain, and now you feel a bit of a sniffle coming on. What can you do to ward off the cold? Common wisdom would suggest that you drink plenty of orange juice. It contains vitamin C and, as everyone knows, this combats a cold.

Vitamin C was discovered in the early 20th century. Almost immediately, it was touted as a cure for colds. Its high concentration in certain fruits chimed in with traditional nutritional advice about consuming more juicy stuff when ill. By the mid-20th century, the link was firmly entrenched in the popular imagination. A trawl through old newspapers brings up endless examples, such as the 'Vitamin C Pudding' recipe in a 1954 Scottish paper, or adverts for the (still popular) blackcurrant drink Ribena, which marketed itself as a cold cure.

Still, the science behind the link was a bit iffy. Some studies had shown a limited benefit to taking vitamin C during a cold, but others reported no effect. Then, in 1970, one of the world's leading scientists intervened. Linus Pauling (1901–94), a double Nobel Prize winner, published a book called *Vitamin C and the Common Cold*. This and subsequent publications made near-miraculous claims for the vitamin. Not only could it ease a cold, but it might also help with heart disease and cancer. Pauling's findings were dismissed by the wider scientific community, but his high-profile campaign had only reinforced the connection in people's minds.

Much research has now been undertaken on the relationship between vitamin C and illness, including the common cold. The results remain inconclusive. The largest study (Hemilä and Chalker, 2013, doi:10.1002/14651858.CD000980.pub4) combined and examined the results from 29 previously reported trials, collectively involving more than 11,000 people. The authors found no link. Vitamin C did not ease the severity or duration of a cold, nor the chances of catching one. However, those undergoing short stresses, such as marathon runners and people exposed to extreme cold, might see some benefit.

If you do seek more vitamin C, then oranges are not necessarily the best foodstuff to go for. A typical serving of broccoli packs in more, without the sugary side-issue. Peppers, cauliflower and sprouts are similarly loaded.

One orange-juice incident that sounds like a myth, but isn't, concerns the 'Sunny Delight' scandal of 1999. A four-year-old girl, it was reported, had drank so much of the processed juice drink that her skin had turned orange. Doctors verified the story. The girl had taken in 1.5 litres (2.6 pints) a day, far more than the recommended limit. Her organs could not process all of the orange beta-carotene pigment, and so it accumulated in her skin. The incident caused red faces among the executives at Sunny Delight, who pointed out that a similar intake of carrot juice would have done the same.

# Tapeworms can help you lose weight

Imagine if you could eat anything you fancied without gaining weight. That's the allure – and surely the only allure – of the tapeworm diet. Adherents swallow tapeworm eggs, which develop into full-on parasites in the intestines. The beasts, up to 17m (55¾ft) long, absorb calories that would otherwise burden the host. It's like having a long, thin safety net inside your alimentary canal, which intercepts food before it can add to your waistline.

The reality is far from inspiring. For starters, you're not going to notice any weight loss when you stand on the scales. The tapeworm might be eating some of your food, but it will itself contribute to your measurable scale reading. You're not losing weight, you're just storing it in a vermicular larder. The worm is also likely to increase your appetite. Some subjects with tapeworms have put on weight thanks to the constant urge to feed the monster. Otherwise, tapeworms tend to cause few, if any, side-effects. Most hosts aren't aware they have one until a section of the parasite breaks off and wriggles out of their anus (and, yes, that really does happen). Those already on a poor diet may experience problems as the parasite laps up nutrients they can ill-afford to lose.

People who swallow eggs from the pork tapeworm face a more sinister risk. The larvae may burrow out of the intestines and get into the bloodstream, from where they can reach other parts of the body. The most serious upshot is neurocysticercosis, an unpleasant condition in which the parasite gets into the eye or brain. Lethal seizures can ensue. Do not Google any of this if you're having tagliatelle for dinner.

# Smallpox has been entirely eradicated

The defeat of smallpox is one of the greatest achievements of medical science. The disease has haunted our species for thousands of years. During the 18th century, an estimated 400,000 Europeans died from smallpox each year. Many survivors were disfigured or blind, or both. A face covered in smallpox bumps is shocking to behold – run an image search with caution.

Smallpox remained a big killer well into the 20th century. As late as the 1960s, an estimated two million people a year died from the disease – mostly in Africa and India. Then came the great push. From 1967, the World Health Organization intensified its vaccination programme. The Herculean effort spanned war zones and remote areas across the globe, and all at the height of the Cold War. Within a decade, smallpox was quashed.

Uniquely, we know the name of the last person ever to contract the disease naturally. Ali Maow Maalin of Somalia was diagnosed with smallpox in October 1977. He made a full recovery and was inspired to work on the eradication of polio in his country three decades later. Sadly, he died of malaria while carrying out the vaccinations.

There have been no reports of smallpox arising naturally since Maalin's case. Tragically, though, a laboratory outbreak caused a death the following year. It happened at the University of Birmingham Medical School in the UK. Medical photographer Janet Parker succumbed to smallpox on 11 September 1978, a few weeks after developing the first symptoms.

It seems that a faulty ventilation duct had allowed virus particles to reach Parker's darkroom. Fortunately, her illness did not spread widely. Only her mother, Hilda Witcomb, contracted the disease, but survived. She is the last documented person to suffer smallpox anywhere in the world, and her daughter the last to die from the age-old killer.

Smallpox may now be eradicated in the wild, but it still exists in two research facilities in the USA and Russia. The World Health Organization has repeatedly recommended that all remaining stocks be destroyed, and has repeatedly been rebuffed. The chance that one of history's deadliest killers might escape confinement is small but not impossible.

Other threats lurk in the shadows. It's perfectly possible that further stocks exist, hidden by governments as part of illegal bioweapon research. Perhaps the greatest risk is that someone with evil intent might create the virus from scratch. A modern biotechnology lab would be up to the task using small DNA building blocks.

One scenario sounds like something from *Game of Thrones*. Across the lands of the north, and particularly in Siberia, centuries-old permafrost has started to melt under warmer conditions. Corpses buried long ago are re-appearing. Some bear the marks of smallpox. The live virus has not been detected, but that doesn't mean it's impossible. As if global warming did not already present the world with enough problems, it might yet unleash an ancient plague upon the world. Speaking of which ...

# 'Ring a Ring o' Roses' is all about the Black Death

Like most folk songs, this old ditty comes in many variations. The most common in Britain goes like this:

Ring-a-ring o' roses,
A pocket full of posies,
A-tishoo! A-tishoo!
We all fall down.

It's often said that the rhyme dates back to the 14th century, when bubonic plague killed something like 60 per cent of Europe's population. The lyrics allude to some kind of illness, while the accompanying dance is truly infectious. Everyone holds hands and parades in a circle. Sneezes ensue, then the group falls to the floor in a playground apocalypse.

Besides the sneezing and collapsing, the song makes other hints at bubonic plague. The ring o' roses can be likened to the red rash around buboes – a symptom of the black death. The pocket full of posies, meanwhile, recalls the herbal preparations carried about to ward off bad smells and – so people thought – the disease itself.

It's a convincing just-so story – but flawed. The dance was not recorded until 1855, when the *Brooklyn Eagle* refers to a hand-holding game called Ring o' Roses. The lyrics were not printed until a decade after that. No published version bears a sneeze until the 1880s, and then only in one of numerous variations. Most tellingly, nobody seems to have interpreted the song as a metaphor for plague until after the Second World War. In other words, the idea that 'Ring a ring o' Roses' is an ancient plague song was invented within living memory.

Speaking of sneezing, it's untrue that your eyes would pop out if you managed to sneeze with your eyelids open. The eyelids shut as part of a general pattern of muscle tightening across the face. Some people are able to override the reflex, and any of us could attempt to hold an eye open while sneezing. In neither case will the pressure from the sneeze reach the eye socket, causing a pop. It is, however, possible for an eyeball to become divorced from its usual home during a trauma or disease. In 2017, US basketball player Akil Mitchell sustained such an injury during a collision. The displaced eyeball can be seen in graphic footage, available on the web if you're so inclined. Following treatment, Mitchell made a full recovery. Others can partially extrude their eyeballs at will. Antonio Popeye (that may not be his real name) demonstrated the ability on TV show *Britain's Got Talent* in 2011.

# Watching lots of TV is bad for your eyes

There may be any number of reasons to discourage a youngster from watching too much television, but concern about their eyesight should not be one of them. There is no evidence that looking at a television screen can cause any kind of eye problem. The same goes for computer screens or phone screens. Basically, almost everything you ever heard about screens causing eye damage is untrue or exaggerated.

It doesn't matter if you sit right in front of the screen, or far away. Whether you opt for plasma, LCD or old-fashioned cathode-ray makes no difference. You can watch for hours on end, at an oblique angle, or even upside down if you like. No amount of exposure is going to cause long-term harm to your vision.

That's not to say a prolonged bout of television is harmless. Hours of goggleboxing can take their toll in other ways. Nobody ever watched all the James Bond films in one sitting and walked away with a skip in their step. You might suffer a headache or drowsiness. Your general health will take a spanking for every day spent lounging on the couch. Hours in front of the TV might even cause eye strain or fatigue, but this will quickly diminish with rest.

The myth has no specific origin. It is one of those pieces of common wisdom that parents make up as a way to control their children. 'Stop watching the TV, kiddo. It'll ruin your eyesight,' often means 'Stop watching the TV because I want you out of the house so that I can watch TV.' Over time, the idea is repeated so often that it becomes received wisdom, despite a complete lack of supporting evidence. Such myths are actually quite common. 'Eat up your veg – it'll help you see in the dark.' 'Stop eating so many sweets or your teeth will fall out.' 'You'll catch a cold if you go out without your coat.' Many are found elsewhere in this book. I like to call them FACTOID myths, which stands for Frustrated Adults Con Their Offspring In Desperation. 'Don't sit so close to the screen or you'll ruin your eyesight,' is the ultimate FACTOID.

# Vaccines can cause autism

Vaccines might just be the most important innovation in human history. Untold millions of lives have been saved by inoculation. Smallpox, polio, tetanus and other diseases once killed or crippled indiscriminately. Nobody dies of smallpox any more. Nobody even catches the virus. Thanks to an effective vaccine and a determined eradication programme from the World Health Organization, you're now as likely to be killed by a dinosaur as to die from smallpox*.

Despite their proven power, vaccines have always attracted criticism. This is not surprising. Vaccines give the immune system a weakened virus to play with, so it learns how to fight against the real deal. A healthy person must be deliberately infected, albeit with a sluggish version of the disease. That doesn't seem like a winning idea on the face of it. Then, too, most vaccines are given to small children. Any parent who has held a quivering infant while a needle is inserted into first one arm, then the other, will not quickly forget the experience, even if the child does. It's frightening.

Objections to vaccination are as old as vaccination itself. Early inoculations relied on lymph matter rubbed into scored skin, an off-putting horror to many. Others have refused on grounds of

---

*FOOTNOTE: But see pages 121–122.

religion, mistrust in medicine, or a belief that inoculation simply doesn't work. On the whole, though, vaccines are today widely accepted. Enough people are inoculated to ensure 'herd immunity' in most cases. In other words, some people still get the disease, but those around them are immune, so there is no outbreak.

Vaccines are under constant scrutiny. Are they as effective as possible? Might they be harmful in certain people? Could they be better formulated, or delivered more economically? This is all good and proper. Any widespread public health programme should be monitored from every possible angle. But occasionally, something turns up that undermines public faith in vaccines. A particularly sad example, whose effects are still with us, concerns a perceived link between vaccines and autism.

In 1998, a group of British scientists published a paper about the measles, mumps and rubella (MMR) vaccine. They found that incidents of both bowel disease and autism were higher among those who had received the combined vaccine, than in those who had not. The research was later picked up by the press. Scared parents began to refuse the vaccine. At the peak of the hysteria, only 81 per cent of eligible children in the UK received the jab. Incidents of measles and mumps increased. Children died as a result.

The journalists really should have dug deeper, earlier. It turned out that there were serious shortcomings with the underlying research. The study was tiny and poorly conceived – just 12 children were recruited, and no control group was considered. The findings were frail, and not enough to draw a link between MMR and autism. Yet once public interest was piqued, the press kept hammering away. Nothing shifts newspapers like a good scare story.

Eventually, the truth came out. A frazzled skein of bad science, dodgy ethics and competing interests was gradually disentangled. The researchers had performed invasive and unpleasant procedures on children without the approval of the hospital ethics committee. Data was allegedly changed to support conclusions. One of the lead scientists involved, Dr Andrew Wakefield, stood to gain financially if the MMR vaccine was discredited.

The paper was retracted and declared as fraudulent. Wakefield was charged with over 30 counts of professional misconduct and 'struck off' the medical register, meaning that he could no longer practice medicine in the UK. Subsequent research, including a 2014 review that took account of 1.2 million children, has found no link whatsoever between autism and vaccination.

The furore has since died down, but not vanished. Wakefield continues to promote his ideas, despite overwhelming scientific consensus to the contrary. In the USA, the anti-vaccine movement has been given a fillip by Donald Trump, who has shown sympathy for a connection on several occasions. In science, it is impossible ever to be 100 per cent certain about anything. Nevertheless, the supposed link between autism and MMR vaccination has now been tested thoroughly, with no risk found. The benefits of inoculation against measles, mumps and rubella far outweigh the tiny possibility that scientific consensus is wrong.

# Famous

# bodies

Napoleon wasn't short, Richard III was not a hunchback and
what exactly was going on with David Bowie's eyes?

# Richard III had a disfiguring hunchback and walked with a limp

'That bottled spider, that foul bunch-backed toad!'

William Shakespeare did not paint a rosy picture of the eponymous king in *Richard III*. The ruler is styled as a stooping, ungainly man, not fit for physical pursuits. The famous opening monologue ('Now is the winter of our discontent ...') runs for just 14 lines before the king admits his supposed bodily shortcomings.

But I, that am not shaped for sportive tricks,
Nor made to court an amorous looking glass;
I, that am rudely stamped and want love's majesty
To strut before a wanton ambling nymph;
I, that am curtailed of this fair proportion,
Cheated of feature by dissembling nature,
Deformed, unfinished, sent before my time
Into this breathing world, scarce half made up ...

And so he goes on. The king concludes that, if he can't charm with his beauty, then he means to thrive through cruelty and deceit. Richard is an exceptionally eloquent pantomime villain.

The king may well have been crooked of mind, but not of posture. Not greatly, anyhow. Richard's body was rediscovered in 2012, beneath a car park in Leicester, England. His spine was indeed strongly curved, a condition called scoliosis. The curve would have made Richard a few inches shorter than otherwise, perhaps coupled with a slight stoop – but certainly not a full-on hunchback. Nor would he have limped. The skeleton also showed no signs of a withered arm, an affliction that some biographers have suggested. In well-tailored robes or fine armour, Richard might have looked every inch the sweet prince.

Those biographers, including Shakespeare, may have been influenced by the ruling dynasty. Henry VII, the first Tudor king, had vanquished Richard to end the Wars of the Roses. Richard was the great baddie of the day. It would have been natural for those dramatizing his life to exaggerate his ogre qualities. Shakespeare wrote *Richard III* around 1592, more than a century after the king's death. Nobody alive could possibly remember the king. His life, looks and demise had entered the realms of hearsay and calumny.

It seems that Richard met a brutal end, though not one consistent with a physical weakling. Near-contemporary accounts, backed up by evidence from his skeleton, suggest that he died in a battlefield melee. His skull shows signs of 11 wounds received in the moments before death. Far from a shuffling hunchback, not shaped for sportive tricks, Richard III was the last king of England to die fighting.

# Anne Boleyn had 11 fingers

The six unfortunate wives of Henry VIII led turbulent lives, but none were so troubled as Anne Boleyn. The King annulled the marriage to his first wife to marry Anne, a move that angered the Catholic church and unleashed centuries of religious troubles in his realm. Henry had Anne's head lopped off just three years later. The severed head was Anne's most conspicuous deformity, but those gazing into her coffin might have noticed another. The Queen supposedly bore a sixth finger on her right hand. It was a malformation to complement Henry's Reformation.

About one in every 1,000 babies has bonus toes or fingers – the rate varies among ethnic groups. Most people with polydactyly, as the condition is known, have only one or two rogue digits. The world record is held by Akshat Saxena, born in India in 2010. Saxena had seven fingers on each hand and ten toes on each foot – a total of 34. The supernumerary digits have since been removed.

Anne Boleyn, by contrast, had only one supplementary finger. Or did she? No portraits show it, and no descriptions from her time make mention. The first reference to the abnormality comes in 1586, exactly half a century after Anne's death. According to Nicholas Sanders, 'It is said she had a projecting tooth under the upper lip, and on her right hand six fingers.' Not only was Sanders reporting hearsay ('it is said') long after the event but – as a Catholic priest forced overseas by the fallout from Henry's marriage – he too had an axe to grind* at Anne. His biased writing influenced others, helping to create the myth of the 11-fingered queen.

Anne's bones have since been examined. In 1876, the floor was taken up inside St Peter ad Vincula, the church within the Tower of London. The skeleton of Anne was found to be jumbled up, as though previously disturbed, but there was no hint of additional finger bones.

---

*FOOTNOTE: Pun intended, if only to highlight another misconception about Anne. She was not executed with an axe, but a sword.

# Napoleon was very short

Someone who is short of stature and also short-tempered might be said to have a 'Napoleon complex'. The term can be offensive. It suggests that the person feels the need to pick fights, as though to compensate for an underwhelming physique. It's also historically uninformed, for Napoleon was not as short as commonly believed. At the time of his death, Napoleon was measured a little over 5ft 6in (approx. 1.7m). Today, that would mark him as shorter than most Frenchmen, but not by much. In the early 19th century, such a height was a pinch above average. Napoleon was no shorty.

The myth of the stunted emperor has several possible origins. He may have looked compact on the battlefield when surrounded by his Imperial Guard – often the tallest, most powerful soldiers. He acquired the nickname *le petit caporal* ('the little corporal') early in his career, which was a term of affection rather than a critique of his height. Another factor may have been that the French system of feet and inches was out of kilter with the British yardstick. Napoleon's height of 5ft 6in would equate to 5ft 2in on the French scale. The latter was used in his autopsy.

A final ingredient – perhaps the most important though often overlooked – was the rise of the political cartoon in Britain during the Napoleonic wars. The emperor was mercilessly caricatured, and often shown abbreviated next to his nemesis the Duke of Wellington. Such images would have cemented the idea of a tiny Napoleon, at least among his enemies. A visit to the Duke of Wellington's former home, Apsley House in London, offers an ironic reversal of this height mismatch. Here you will find a naked statue of Napoleon that towers 3.45m (11ft 4in) up a stairwell. It is the ultimate repudiation of *le petit caporal*.

Like many a great figure, Napoleon's demise has always attracted myths and conspiracies. His official cause of death was a cancerous growth in the stomach, but rumours soon circulated that the Emperor had been poisoned. That's not so far-fetched. Given the opportunity, half the population of Europe might have volunteered. Napoleon himself had predicted his death a few weeks before, writing in his will: 'I die before my time, murdered by the English oligarchy and its assassin.' The leading theory was a fatal dose of arsenic. In 1840, when exhumed for reburial in Paris, Napoleon's body was found to be very well preserved. This is consistent with arsenic poisoning. The element is fatal to most bacteria as well as to humans, preserving the tissues after death. A later theory, still commonly believed, is that the arsenic came from his wallpaper. Boney did not fall foul of a murderer, but interior decoration. This idea is now all but debunked. Modern analysis of Napoleon's hair samples shows no evidence of arsenic poisoning (J.T. Hindmarsh and J. Savory, 2008, doi:10.1373/clinchem.2008.117358).

While we're dealing with the era, we might as well turn to another of the Emperor's arch-enemies, the British Admiral Lord Nelson. Nelson has lingered in the popular imagination for longer than most military heroes of the era. Partly, this is down to his heroic demise at the Battle of Trafalgar. The enormous and eponymous column in Trafalgar Square, London makes him hard to forget. Still, the common image of Nelson is somewhat warped. He's often pictured wearing an eye patch. There was no need. His blind right eye was not missing or even disfigured. Documentation from the time suggests that anyone looking at the Admiral would find it hard to work out which eye was sightless. The patch was invented by portraitists, long after his death, to suggest triumph over adversity. Film and TV dramatizations have reinforced the image. You won't find such rogue accoutrement in any portrait or bust from his lifetime – and that goes, too, for the statue in Trafalgar Square.

Oh, and in case you were wondering, the great warrior of the sea was only 5ft 4in (1.63m) – a little shorter than Napoleon.

# Jeanne Calment was undisputedly the oldest person who ever lived

Jeanne Louise Calment remains the only human to make it into her 13th decade. She was born in 1875 and passed away in 1997 at the age of 122 years and 164 days. She could recall meeting Vincent van Gogh as a teenager, and watched the Eiffel Tower appear on the Paris skyline. Remarkably, she lived alone until the age of 110, and gave up smoking at 117, 95 years after she'd taken up the habit. On her death, she was recognized by Guinness World Records as the oldest person who ever lived.

Calment's lifespan is mind-boggling, yet there may be people alive today who are older. The supercentenarian was fortunate to have been born in a time and place with reliable record keeping. Her age is verifiable from an official birth certificate, and so she is eligible for the record books. There are many others across the world, and throughout history, who may have lived longer but could not prove it.

The Old Testament, of course, is populated with hoary patriarchs. Adam and Noah made it past their 900th birthdays, while the Methuselah of cliché is said to have lived to 969. Other faiths and traditions have their equivalent supercentenarians.

Historical, as opposed to mythical, claimants are also easy to find; 17th century Britain, in particular, enjoyed a craze for the long-lived. The most famous was Thomas Parr, known as Old Parr, who died in 1635 at the supposed age of 152 (some sources say 169). The case was taken so seriously that Parr was painted by the likes of Rubens and van Dyke. He was eventually buried in Westminster Abbey, a distinction afforded to only the most eminent. Other celebrated antiques include Yorkshireman Henry Jenkins, who died in 1670 claiming 169 years; and Chesten Marchant, a Cornish woman who died in 1676 at the age of 164. These and other contenders must be taken with a pinch of salt. In the absence of official record keeping, a person might claim to be 300 and nobody could prove otherwise. That said, any of these claimants might have genuinely reached an age that would have beaten Mme Calment. We will never know.

Even today, there are pretenders to the wizened throne. A Nepalese man, Bir Narayan Chaudhary, was said to be 141 just before his death in 1998. Moloko Temo of South Africa allegedly reached 134 before her demise in 2009. Most recently, Mbah Gotho of Indonesia had notched up a self-reported 146 years by the time of his departure in 2017. Such extreme longevity is unlikely but cannot be ruled out. It is quite possible that someone, somewhere has lived more than the 122 years of Mme Calment but lacked the birth record to prove it.

At the time of writing, the oldest living person in the world is Nabi Tajima of Japan, aged 117. She was born on 4 August 1900, which means she is the last surviving person from the 19th century (as, strictly speaking, new centuries begin in the year '01' and not '00'). In a related feat of longevity, two grandchildren of John Tyler, the tenth US President, are still alive (again, at the time of writing). That sounds impressive enough, and then you see the dates. Tyler was born in 1790 and President from 1841–45. He fathered 15 children, well into his sixties. One of his children carried on the tradition, becoming a father aged 75. The two remaining grandchildren, Lyon Gardiner Tyler and Harrison Ruffin Tyler were born in 1924 and 1928, respectively. They can boast a famous grandfather who was born while George Washington was President.

# The golden Bond girl shows how paint can asphyxiate

'She died of skin suffocation,' says Bond. 'It's been known to happen to cabaret dancers. It's all right so long as you leave a small bare patch at the base of the spine to allow the skin to breathe.'

We round off this section with a fictional body, and one of the most memorable deaths in cinema history. In the third James Bond movie, *Goldfinger* (1964), the titular baddie is betrayed by his secretary Jill Masterson, played by Shirley Eaton. Goldfinger gets his revenge in flamboyant style. The rogue employee is covered in gold paint until she asphyxiates. As Bond (played here by Sean Connery*) later observes, after finding the body, this prevents the skin from breathing. Death is swift.

Everybody knows that we take in oxygen through our lungs. But the skin is another organ with a large surface area exposed to the atmosphere. Can it also absorb oxygen? Can it 'breathe'?

The Discovery Channel show MythBusters made two attempts at re-creating the scene from *Goldfinger*. The hosts were slathered in gold paint and

---

*FOOTNOTE: Little known body oddity – Sean Connery wore a hairpiece in all of his James Bond films, to disguise his receding hairline.

hooked up to monitors. One host experienced a change in blood pressure and body temperature, but the follow-up showed no change whatsoever. Their dignity may have taken a battering, but neither presenter's life was ever in danger. Death by gold paint was branded a myth.

And it is. No human would suffocate if painted in this way. This is obvious when you think about it. How do Scuba divers – including Commander Bond himself – survive? The diver's skin is isolated from oxygen, save for a tiny amount in the facemask that would soon be exhausted. They suffer no ill effects.

You might encounter other problems if you were entirely coated in paint. Sweating would be a challenge. You might overheat. Or you could experience symptoms from any lead or impurities in the paint. Were you to be treated with actual gold, you would burn to a crisp – the elemental metal has a melting point of 1,062°C (1,944°F). But you would not suffocate.

The idea that the skin breathes goes back to at least the mid-19th century and remained scientific orthodoxy at the time of *Goldfinger*. Indeed, medics were on hand during the filming. Eaton was not covered completely in paint, but nobody wanted to take any chances. She suffered no ill effects, though a daft urban myth – easily dispelled by looking at her IMDb record – would have us believe that she died during filming.

Medical science has since shown that all the oxygen we need is taken in through the lungs*. The outer layers of skin can absorb oxygen from the air, but its contribution to our wellbeing is nil to negligible. Feel free to gild yourself whenever you like.

---

*FOOTNOTE: There is one exception. The cornea of the eye must be transparent to let in light. It perforce lacks blood vessels, and so oxygen cannot be supplied in the usual way. The cornea gets around this by taking in oxygen dissolved in water (tears) or directly from the atmosphere through absorption.

Ian Fleming, author of the Bond novels, seems to have been fixated with body oddities. Dr No has prosthetic hands. Arch-nemesis Ernst Stavro Blofeld bears a deep scar above and below his right eye, while Emilio Largo went for a villainous eye patch. You might also recall the curious case of Francisco Scaramanga in *The Man with the Golden Gun* (played by Christopher Lee in the movie). Scaramanga's peculiarity, besides a passion for ludicrous weaponry, was the possession of a third nipple. This condition is a medical reality. Actor Mark Wahlberg is perhaps the most famous celebrity with a supplementary nipple. These are an evolutionary throwback; an echo of an ancestor species that had litter births and needed multiple suckling options. Curiously, Scaramanga's third nipple rides high on his chest, in a position where nipples have no business.

A final Bond body myth was started by the novel and film *You Only Live Twice*. Here, Bond is told that sumo wrestlers are able to retract their testicles for protection during combat. The story can be dismissed with an apposite word that rhymes with pollocks.

# And a few more odd bods ...

**David Bowie** did not have eyes of different colours. The pupil in his left eye was permanently dilated – in other words, it was much bigger than the pupil in his right eye. This mismatch made the left eye look naturally darker. It was also more prone to red-eye effects, when light bounces off the back of the eye. Combined, these gave the illusion that the two eyes had different coloured irises, but in fact both were blue. Celebrities who genuinely do have different coloured eyes – or at least partially – include Jane Seymour and Superman actor Henry Cavill.

**Walt Disney** did not have his head cryogenically frozen after death. The cartoonist was cremated on 15 December 1966. Disney is thought to have expressed an interest in cryogenics, but the procedure never troubled his last will. The urban legend arose about a decade later but remains stubbornly persistent. Ironically, the highest grossing animated film of all time is Disney's *Frozen*.

**Adolf Hitler** may or may not have had two testicles. A British propaganda song from the Second World War opens with the immortal lines 'Hitler has only got one ball, the other is in the Albert Hall'. That second part is untrue (I've asked), but it's harder to discount the missing testicle. Rival accounts suggest that Hitler lost a gonad during the Battle of the Somme, or that he bore a condition called cryptorchidism, when only one testicle descends into the scrotal sack. Medical documents are said to support each claim, although nothing has ever been scanned and published for all to see. We can probably rule out the explanation from the song, which holds that 'his mother (the dirty bugger) cut it off when he was small.'

**Elvis Presley** was blond. The King's trademark raven locks were achieved with hair dye. His natural colour was sandy-brown, as can be seen in photographs from his youth. Elvis began dying his hair in his teens and found that the black look fitted his (at the time) edgy image.

**George Washington** did not have wooden teeth. The first US President, like many people from his era, had rotten teeth and commonly wore dentures. These were made from a variety of materials, including gold, ivory, brass, other people's teeth and (living dangerously) lead. No set of wooden dentures is known. It's unclear where the myth comes from. One theory is that his ivory dentures got so stained with wine that they took on the appearance of wood.

**Mel Blanc** was not allergic to carrots. It would have been deeply ironic if the voice behind Bugs Bunny was unable to eat his character's favourite snack. The false rumour got about because Blanc would routinely spit out the vegetable during recording sessions. He couldn't possibly eat every carrot he was asked to crunch while voicing the bunny, so he simply used a spittoon.

# Are you pronouncing it wrong?

Most branches of science contain tricky, technical words. We can ordinarily ignore them. Few need to be fluent in the twists and turns of astrophysics or the ponderous names of chemical compounds. Medical science is different. We all have bodies, and we're all obsessed with their good health. Here, then, are a few of the more complicated words from anatomy and medicine, and how not to be wrong when pronouncing them.

**Cirrhosis:** This scarring of the liver has an unusual name, made up in 1819 by René Laënnec. It comes from the Greek word for tawny or yellowish and is pronounced 'se-rho-sis'. Incidentally, Laënnec also invented and named the stethoscope, and coined the word melanoma.

**Coccyx:** As difficult to spell as pronounce, the coccyx is the fused vertebrae at the bottom of the spine and a remnant of an ancestral tail. It is pronounced 'cock-six'. Curiously, this peculiar word is derived from the Greek word for cuckoo. The human coccyx was thought by Ancient Greek writers to be shaped like a cuckoo's beak.

**Diphtheria:** The temptation here is to take that first three letters and run with them. This now rare bacterial disease is commonly pronounced as 'dip-theria', but note the additional 'h' after the 'p': it should properly be 'diff-theria'.

**Down/Down's syndrome:** This genetic disorder is easy enough to pronounce, but should there or should there not be a possessive apostrophe? It's a thorny question in medicine, and also applies to

Parkinson's disease, Alzheimer's disease, Huntington's disease and dozens of other conditions. There have been movements to drop the possessive. Those meddlesome apostrophes could be misconstrued. We might assume that Down, Parkinson and Alzheimer were all sufferers of their eponymous conditions. They get possessive apostrophes because they possessed these diseases. Not so. These are the names of the doctors who first put down a description. They neither owned nor possessed the diseases, so the apostrophes are misguided.

The notion has found little traction. Down syndrome seems to be more common than Down's syndrome – perhaps because the possessive 's' is effectively silent. Alzheimer's disease, on the other hand, tends to cling on to its punctuated ending. It just sounds wrong without it. In truth, there seems to be little consistency. The only exception comes for those diseases named after the people who had them, which always keep the apostrophe. Lou Gehrig's disease (as ALS is often known in the USA) is the most common example.

**Infarction:** Myocardial infarction is the medical term for a heart attack, when blood flow to part of the heart is reduced or stopped. Infarction refers to the tissue death that results. It's not a word most of us encounter in other contexts, and so the similar and more common word 'infraction', is often used in error. Myocardial infraction would imply that your heart tissue had committed some kind of parking offence or similar. Even medical journals occasionally make the mistake, possibly down to limitations of spellcheck software.

**Mastectomy:** The partial or complete removal of the breast, usually in defiance of cancer. As medical terms go, the word is very well-known due to the prevalence of the disease. It is, however, often mispronounced as 'mass-ectomy' without the first 't'.

**Migraine:** Do you say 'my-grain' or 'me-grain'? Neither is incorrect. The former is favoured in North American English, whereas both forms are used in the UK. The word ultimately derives from the Greek *hemikrania* (half the skull), alluding to the common symptom of pain in one half of the head. Given that 'hemi' is pronounced 'hem-me', we might argue in favour of 'me-grain'. Even so, those who pronounce it as 'my-grain' will always think that the 'me-grain' folk are a bit funny (and vice versa).

**Microchimerism:** We met this phenomenon in the DNA section on page 56. It is the process by which genetically distinct cells from your parents and other relatives end up within your body. At first blush, the word looks complicated and sciency, but it's a doddle to pronounce if you know your ancient mythology. The name comes from the chimera, an ungainly beast that was part lion, part goat and part snake. Being of Greek origin, the 'chi' has a hard sound, as in kayak. So, the whole word is pronounced 'micro-ki-mer-ism'.

**Ophthalmologist:** A doctor specializing in the diagnosis and treatment of problems with the eye. Ophthalmologists are doubly dishonoured: not only are they commonly mistaken for optometrists or opticians (who typically work on the high street, testing vision and recommending corrective lenses, respectively), but many people also pronounce the word incorrectly. 'Op-thal-mologist' isn't quite right, because there's a pesky extra 'h' after that 'op'. The word is more correctly pronounced as 'off-thal-mologist'.

**Prescription:** Often muddled with the word proscription. Prescription is an act of authorization (allowing a patient to access medicine), while proscription is the act of forbidding. They are almost opposites.

# Other myths, misconceptions and misnomers

**Alternative medicine:** The alternative to medicine is 'no medicine'. A medicine is something with the power to cure, alleviate or prevent disease. If the intervention can be shown to do one of these things, then it is medicine. If it cannot, then it is not a medicine. There is no middle ground. There is no third 'alternative'. Compounds can vary in their effectiveness, but if they show no effect whatsoever, then they are not medicines. Items commonly called 'alternative medicines', such as homeopathic treatments and acupuncture, have shown no measurable effect (beyond placebo) when put to rigorous scientific test.

**'Atchoo!':** In English-speaking countries, the human sneeze is typically written as shown here. 'Atchoo!' is a fair reflection of the sound of a sneeze to many of us, but it's not universal. The 'Ahhh' sound at the beginning is inevitable, as we take a deep breath in anticipation of the sneeze. The 'choo' bit, though, is cultural. The French often enunciate an 'achoom', while the Japanese go for 'akashun'. We pick the culturally acceptable sneeze sound up from those around us as young children. Infants and deaf people often make less vocalized sneezes, simply expelling air with no sound effects.

**Boner:** The human penis, at its happiest, is a sturdy feature. Its rigour is achieved through fluid pressure, as the member fills with blood to achieve an erection. There is no bone in there, as is abundantly clear from a fiddle with the unexcited version. Humans are unusual in their flexible members.

Many other mammals, including primates, do enjoy the extra support of a penis bone, known as a baculum. Chimps have it. Bats have it. Grizzly bears, seals and cats have it. The walrus is king dong. One baculum specimen from an extinct species measured 1.4m (4ft 6in) – three times longer than any bone in the human body. A human 'boner', meanwhile, contains no bone, but nobody's going to call it a hydrostat.

**Cervix:** The term is most commonly associated with the muscle-lined entrance to the womb from the vagina. It is the cervix that is screened for abnormal cells during a smear test. It is not, however, the only cervix in the body. Medically, the term cervix describes any narrowed region. The neck of the bladder is sometimes referred to as a cervix, as is the part of a tooth that slims down as it enters the gum. The neck itself is sometimes described by its Latin name, as in the case of a cervical rib (see page 26).

**Chickenpox:** A viral disease associated with a nasty all-over rash. Unlike cowpox, which does affect cows, chickenpox has no known links to fowl. There are several theories as to where the name came from. The most likely is that this is a relatively mild form of pox, and chickens are seen as lowly animals. Or perhaps chickenpox is a corruption of child's pox, this being a disease that is most common in children. An unlikely explanation plays on the resemblance of the scabs to chickpeas. It may even have come from an Indian coin called a chickeen, whose low denomination became a metaphor for low-stakes gambling and therefore, perhaps, diseases.

**Digestive biscuits:** A mainstay of the British tea break, digestive biscuits do not aid digestion in any way. Even so, the brown snacks were developed in the 19th century to help with stomach pains. Digestives contain plenty of baking powder, which can work as an antacid, relieving the symptoms of heartburn (see below). Sadly, the chemical is altered during the baking process, and the completed biscuit has no beneficial effects. They're good for dunking in tea, though.

**Double jointed:** Can you bend your fingers backwards, or touch your chin with your elbow? Those who can pull off such feats are often described as double-jointed, as if they possessed an extra hinge somewhere. It's a misnomer, of course. Those with 'double-joints' have single joints like the rest of us, but may have particularly pliable connective tissue, or bones with unusually shaped ends. Being double-jointed can lead to some amusing party tricks, but it's not a skill for which I'd bend over backwards.

**Dry as a bone:** Skeletons peeping out of the desert sands in a Western film might be a good simile for dryness, but the bones inside you are rather damp. Living human bones are anywhere between 20 and 50 per cent water (depending how you measure). They also contain blood vessels (full of very wet blood), and bone marrow (soft, spongy and, again, pretty moist).

**Elbow licking:** A spurious anecdote says that it's impossible to lick your own elbow, and that 80 per cent of people who are told this fact will immediately try. I know I just did. While the manoeuvre is beyond the talents of most, there are plenty of videos and photos online to prove that it is not a universal handicap.

**Funny bone:** That strange sensation when we jolt an elbow is not caused by a bone, funny or otherwise, but a nerve. Specifically, it is the ulnar nerve, which runs all the way down to the little finger and its neighbour and is close to the surface at the elbow. A bash to the nerve compresses it against the humerus bone which, as punsters will quickly spot, is indeed a funny name.

**Heartburn:** This unpleasant feeling in the upper abdomen is caused by gastric acid from the stomach intruding into the oesophagus. It has absolutely nothing to do with the heart and does not cause burns – though the pain can be felt in part of the chest close to the heart. Indeed, the symptoms can feel similar to the chest pains associated with heart disease, as the oesophagus and heart are wired up to similar nerves.

**Junk DNA:** Genes – the 'useful' bits of DNA that code for proteins – make up just 3 per cent of the genome. The remaining 97 per cent is often described as 'junk DNA', implying that it is without purpose. I'm not sure any scientist has believed this for a long time. At the very least, the superfluous DNA must surely play a role in folding and packing the genes into a favourable conformation. It's now clear that so-called junk DNA does much more. Some sections, for example, act as switches for gene expression – the complicated and beautiful dance by which a gene's instructions are read and used as a recipe for building a protein. No doubt other roles will be found in time. The term 'junk DNA', once a mainstream scientific phrase, is now rarely used by scientists, but still crops up in more colloquial discussions.

**Knuckle cracking:** It's often said that cracking one's knuckles is a sure way to develop arthritis. It has a certain logic. If you put pressure on your joints habitually over a lifetime, then you're likely to do some damage. The premise has, in fact, been tested scientifically (R.L. Swezey, S.E. Swezey, 1975, *Western Journal of Medicine*, 122(5):377-9, no doi). Researchers found no link between life-long knuckle cracking and a higher incidence of arthritis. The group size of just 28 people of the same ethnicity could hardly be called conclusive, but no other evidence has emerged to link the annoying habit with arthritis.

**Malaria:** This unpleasant infectious disease has always plagued our species but was only given its modern English name in the 19th century. It was long believed to be caused by miasma, a disease-carrying vapour. Indeed, malaria gets its modern name from the medieval Italian phrase mala aria, which means 'bad air'. It was discovered in the early 20th century that the disease is caused by Plasmodium parasites spread by mosquitoes, and not the pungent odours associated with their swamp habitats.

**Schizophrenia:** Like many mental disorders, schizophrenia can encompass a raft of different symptoms and behaviours. Sufferers may act antisocially, hear voices that aren't there, or have difficulty sifting reality from imagination, among other things. The term is often applied, mistakenly, to suggest someone with a 'split personality' like Dr Jekyll and Mr Hyde. This condition is known as dissociative identity disorder (DID). Those with DID slip between two or more distinct personalities, often with little memory of their other self or selves. It's not impossible for schizophrenics to have DID as well, but it is uncommon and certainly not a key hallmark of the disorder.

**Shatner's Bassoon:** The British satirical show *Brass Eye* (1997) was renowned for exposing the ignorance of those in power. In one show, a series of celebrities were duped into speaking out against 'Cake', a drug invented by the show's creators. Cake apparently affected Shatner's Bassoon, the area of the brain concerned with perception of time. Bernard Manning, a crass comedian, even fell for the idea that one girl had 'thrown up her own pelvis' after taking the drug. The ruse was so well executed that a Member of Parliament raised the issue in the House of Commons. Needless to say, there is no part of the brain called Shatner's Bassoon. The phrase sounds simultaneously ludicrous and plausible. After all, the body contains such unlikely structures as Hesselbach's triangle, the Golgi apparatus, the pouch of Douglas, the crypts of Luschka, the loop of Henle and – my favourite – the bundle of His.

**Signs and symptoms:** This sounds like a tautology, but the two terms have different meanings. A symptom is subjective, while a sign is objective. What does that mean? Well, if you're feeling a bit groggy then your symptom is nausea. It is something only you are aware of, which can't be detected by anyone else. If you then vomit all over someone's shoes, then you've kindly provided objective evidence that you are sick. The vomiting isn't a symptom but a sign. For this reason, doctors don't usually speak of symptoms when treating a baby. The infant may itself feel symptoms, but can't report them, so only signs (measurable, objective criteria such as temperature) are available for diagnosis.

**Slipped disc:** The 24 vertebrae of the spine are separated from one another by shock absorbers known as discs. Like a satsuma, the disc consists of a soft inner part surrounded by a fibrous coating. Sometimes, the coating gets torn, allowing the softer material to bulge out. The result can be excruciating. The condition is commonly known as a slipped disc, but the term is misleading. Discs are tightly attached to the vertebrae and cannot physically slip in any way. The medical terminology for this affliction is a spinal disc herniation.

**Test-tube baby:** Louise Brown was the first human conceived by in-vitro fertilization, or IVF. During this now-routine procedure, egg and sperm are brought together outside the body. The fertilized egg is then implanted into the womb. Children created in this way were quickly dubbed 'test-tube babies', but the term is erroneous. The fertilization typically takes place in a petri dish rather than a tube. The term in vitro is common in biology and medicine. It is Latin for 'in glass', implying a tube or dish, but is used more widely to describe any biological procedure performed outside the body. This is distinct from an in-vivo procedure, which takes place inside the body.

**Vitamin D:** Strictly speaking, a vitamin is an organic compound necessary for health that is not synthesized in the body and must be gained through diet. Vitamin D is an oddity in this respect. It's readily made in the body so long as that body is getting plenty of sunlight. Those with darker skin are at greater risk of deficiency. The sun's rays are partly blocked by the melanin pigment, reducing vitamin D production. Most people get enough from foods such as eggs, mushrooms and fish, however.

# Let's make up a new wave of false facts

Having debunked so many myths about the human body, we need to invent some replacements. Let's try these.

The back of the eyeball looks a lot like the front of the eyeball. A faded iris and pupil are imprinted there. It is a relic from millions of years ago, when an ancestor species had the ability to swivel its eye 180 degrees if damaged.

Samuel Morse developed the eponymous code after pondering his own arm. He deliberately devised the code for 'SOS' to match a series of shrapnel scars and freckles along his wrist.

Everyone's heard of ingrown toenails, but it's also possible (though rare) to suffer from ingrown teeth. In one extreme example, a Texan man's upper incisors went so far the wrong way that they could be seen through his nostrils.

The human chin once extended out several inches. The pointy prominence can be seen on many Egyptian sarcophagi, and lives on in folk memories of witches and leprechauns. The bones have since evolved into a squarer jaw.

People with ginger hair are more likely to sneeze in threes, and nobody knows why.

Billy the Kid was so named because he possessed a pair of miniature horns like a young goat. This is why he always wore a hat.

If you laid all of your bones end to end, they would stretch as high as the Eiffel Tower.

In the rare condition known as hoot-neck, the bones of the upper vertebrae fuse together, effectively forming a ball-and-socket joint with the skull. Sufferers are able to rotate their heads 360 degrees, like an owl.

The metal-clawed character of Wolverine from the X-Men films and comics is partly based on medical reality. A rare disease, called ferrodigititis, leads to an accumulation of iron in the fingernails. In extreme cases, the iron can combine with carbon and calcium deposits to form steel-like claws.

The longest tapeworm on record measured over a quarter of a mile in length. Its body mass was greater than that of the patient from whom it was removed. Technically, this is a case of a human parasite on a worm host.

Three percent of the population has olfactory bulbs in their ears. The surest way to tell is if you think you can 'smell' chlorine while swimming underwater in a pool – even though you can't sniff.

It's well known that humans share many genes with our great ape cousins like the chimps. Our so-called 'junk DNA', though, is a closer match to that of the squirrel. Scientists have no idea why.

# Index

# Other titles in the series

Everything You Know About
London is Wrong
9781849943604

Everything You Know About
Science is Wrong
9781849944021

*"A great read and a top present
for any smarty pants"*

The Sun

All published by Batsford.

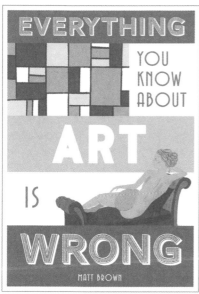 

Everything You Know About          Everything You Know About
Art is Wrong                        Space is Wrong
9781849944298                       9781849944304

*"Crammed full of quirky facts,
it is hard to put down"*

This England

# About the author

Matt Brown holds degrees in Chemistry (BSc) and Biomolecular Science (MRes). He has served as a scientific editor and writer at both Reed Elsevier and Nature Publishing Group. He served as the Royal Institution's quizmaster for several years, and has also put on science quizzes for the Royal Society, Manchester Science Museum, STEMPRA and the Hunterian Museum in London. He is author of five previous books, including other volumes in this series on London, Science, Art and Space – all available from Batsford. Find him on Twitter as @mattfromlondon, and follow the series on @eykiswrong.